45 Topics in Current Chemistry
Fortschritte der chemischen Forschung

Dynamic Chemistry

Springer-Verlag Berlin Heidelberg GmbH 1974

This series presents critical reviews of the present position and future trends in modern chemical research. It is addressed to all research and industrial chemists who wish to keep abreast of advances in their subject.

As a rule, contributions are specially commissioned. The editors and publishers will, however, always be pleased to receive suggestions and supplementary information. Papers are accepted for "Topics in Current Chemistry" in either German or English.

Any volume of the series may be purchased separately.

ISBN 978-3-662-15518-9   ISBN 978-3-540-37787-0 (eBook)
DOI 10.1007/978-3-540-37787-0

Contents

Empirical Force Field Calculations

A Tool in Structural Organic Chemistry

Dr. Cornelis Altona and Dr. Dirk H. Faber

Gorlaeus Laboratory, The University, Leiden, The Netherlands

Contents

I. Introduction

Every student and researcher working in organic and structural chemistry nowadays feels free to use mechanical molecular structure models in order to gain insight into the configurational or conformational properties of the molecules under study, to explain the results of physical measurements, to discuss even fine points on chemical reactivity at different sites in a molecule. This practice is currently accepted in the chemical literature, but it should be remembered that "thinking in three dimensions" emerged only recently in chemical history. Mohr's discussion [1] on structural properties of cyclo-hexanoid systems (including bicyclo[3.3.1] nonane and adamantane!), illustrated by ball-and-stick drawings, unfortunately does not seem to have made a favourable impression on his contemporaries. For example, the first three-dimensional representation of the skeleton of norbornane appeared as late as 1935 [2], the first distinction between axial and equatorial positions on a cyclohexane ring was made in 1931 [3]. Seen in historical perspective, important advances in chemical thinking have always been accompanied by a consistent and daring use of certain *models* as a basis for further experiments and for building up our theoretical insight. For example, Barton [4] opened new ways in chemical thinking by showing the existence of meaningful correlations between chemical reactivity on the one hand and the position of the substituent (axial *vs* equatorial) on the six-membered rings of the steroid skeleton assuming the "all-chair" model. In the decade following Barton's publication numerous papers have appeared in which the course and rates of reactions involving certain fundamental groups in saturated non-aromatic systems were related, albeit often more qualitatively than quantitatively, to structural features at the reaction site. Several specialized textbooks have treated these subjects exhaustively. Two more or less interrelated concepts were (and are) especially useful tools for the organic chemist, one concerning the purely stereochemical requirements ("geometry") of the molecule before, during, and after undergoing a chemical reaction, one concerning the energetic requirements ("strain"). During the past few decades the use of mechanical models of organic molecules, endless variations on the ball-and-stick system, has been commonplace in chemical laboratories, though the limitations inherent to a merely mechanical representation have not always been clearly understood. That is to say, chemists are sometimes tempted to transfer properties peculiar to the hardware of the model to properties of the molecule under study. An interesting example of pitfalls sometimes encountered in this area is furnished by the unexpected geometry of some $9\beta,10\alpha$-retrosteroids. In one instance the solution of the crystal structure problem was unsuccesfully tackled with search models based on Dreiding scale models [5]. These models suggested a regular chair conformation of ring C in the $9\beta,10\alpha$-pregna-4,6-diene-3,20-dione system and an overall

L-shape of the steroid skeleton. The geometry actually found is a flattened chair for ring C [5,6] and a nearly planar skeleton!

Nowadays an *a priori* computer calculation of the energy and detailed geometry of molecules the size of steroids is quite feasible [7-9] and can be regarded as a practical research tool. An outline of the scope and present-day limitations of this tool, alternatively known as Westheimer method, strain energy minimization technique, molecular mechanics or force-field method, will be the subject of this review. We wish to make it clear at the outset that the output of a force-field (FF) calculation of a given molecule (consisting of *a priori* calculated geometry, energy and so on) *is to be regarded strictly as a model*, albeit a model which is vastly superior to hardware models as was already pointed out by Hendrickson [10] in his pioneering paper on computer calculation of hydrocarbon rings.

Computer calculations

Roughly three types of approaches to calculations of molecular properties may be distinguished[11], each of which is important in its own area of applicability.

(1) Strictly *ab-initio* SCF-LCAO-MO calculations, in which no empirical parameters are required.

(2) Semi-empirical (simplified) quantum mechanical treatments at various levels of sophistication (EHT [12], CNDO [13], MINDO [14], NDDO [15], PCILO [16], as well as others).

(3) Phenomenological treatments which approximate the molecular potential field (Born-Oppenheimer approximation) by a series of "classical" energy equations and adjustable parameters. These treatments may be called "classical mechanical" only in the sense that harmonic force-field expressions stemming from vibrational analysis methods are often introduced, though strictly speaking one is free to select any set of functions that reproduces the experimental data whitin chosen limits of accuracy.

Methods of type (1) are to be used when the fundamental physical origin of phenomena, such as potential barriers or the "anomeric effect" for example, are the object of study. Lengthy calculations are the price one has to pay for quantitative accuracy. At present, prohibitive amounts of computer time would be needed if one wanted to do an accurate *ab-initio* study of the conformations of a sizeable organic molecule, even if the precise molecular geometry is known. Still, one may hope that *ab-initio* calculations with full relaxation of internal coordinates, carried out on a selected series of small molecules, will in the near future provide workers employing methods (2) and/or (3) with a better basis from which to start operations. Of special interest in this respect are calculations aimed at the elucidation of the intramolecular potential field as function of the deformation of internal coordinates.

Method (2), simplified quantum mechanical treatment, also has its strong points and drawbacks. The single great advantage is that relatively few parameters are required to calculate energies and other desired properties of any given organic molecule of *known geometry* (within practical limits of size). We feel that in cases where the geometry is not known with confidence from physical measurements or from calculations by method (3) the results of method (2) should be viewed with some caution, particularly when the molecule is severely "strained". Experience has taught that a complex organic structure has many devious and sometimes unforseeable ways of partitioning strain over its internal coordinates. Therefore, quantum mechanical calculations of molecular energy that are based on "standard" bond lengths and valency angles or on any guessed-at geometry may be in error by several kcal/mole from this source of error alone. Prospects are better when the molecular geometry is optimized during the calculations.

Our discussion from now on will be limited to method (3) which is *practically useful* in calculation of accurate geometry and other properties of complex molecular systems. One may obtain quantitative information with a certain degree of confidence: physical accuracy in terms of fundamental physical processes cannot be required. These treatments rely heavily on the availability of at least some experimental data (geometry, conformational energy, vibration spectrum) for the systems under study. By this we mean that the force-field approximations are best regarded as extrapolations, If, for example, the possible conformations of a complicated hydrocarbon molecule are to be predicted, one starts off with calculations of a selected series of well-known small molecules, like simple open-chain aliphatics and/or cyclic systems, matches calculated properties against observed ones, then transfers the best forcefield obtained to the molecules of interest. Parameters for any choosen functional group may be added the same way. Surprisingly accurate results are obtained in many cases by this theoretically crude method. Of even greater interest are "failures", systematic discrepancies between observed and calculated properties, because these point to weaknesses in the forcefield used and open the way for improvement. No doubt the parametrization required is the most laborious part of the force field approach, and most workers in the area prefer to adopt one of the force fields extant in the literature.

For practical purposes it is not necessary to be familiar with the details of the basic derivation and mathematical aspects of the forcefields calculations. Nevertheless, even the practical user of the method should not disdain the benefits derived from understanding its theoretical justification. First, it gives insight into the construction of simple phenomenological models for the force field. Second, it provides the practical user on the one hand with a feeling of confidence in the validity of his results, on the other hand with the necessary caution concerning the physical interpretation of the computer output, which should not be stretched too much.

Summarizing, the purpose of this review is not to give a more or less complete coverage of the literature, but rather to illuminate the present scope and limitations of the force-field method and to draw guidelines for the future. We prefer to concentrate our attention mainly on theoretical results, on open questions, and on selected applications taken from recent work of leading experts in the area. A larger body of older but still relevant literature on the FF method and related topics has been covered in an extensive review by Williams *et al.* [17]

The large body of literature on calculations of structures of proteins, polypeptides, polysacharides and polynucleotides is not reviewed. These calculations, in so far as they apply to the force-field approach, are necessarily based on highly simplified fields (hard- or soft sphere approach) and further-more, due to limits in computer memory capacity and speed, full relaxation of the atomic coordinates of such large molecules (> 75 atoms) is as yet unattainable.

II. General Considerations

The Force-Field Geometry and Energy Optimization method (molecular mechanics) views a molecule as a system of particles held together by forces or "interactions". These forces, and the potential energy functions from which they are derived, are for practical reasons split into various components:

(1) Forces involving two atoms; these include stretching of bonds, inter-actions between atoms bound to a common atom (Urey-Bradley forces), and the common non-bonded interactions (van der Waals-London forces).

(2) Forces involving three atoms, *i.e.* bond angle bending.

(3) Forces involving four atoms, *i.e.* torsional forces.

The above forces, represented by their derivatives or diagonal matrix elements, are easily visualized and are assumed to have a certain tangible physical interpretation. One may note that at this point the road of the MMFF method splits off from the traditional road taken by the vibration

5

spectroscopist, though part of the language remains similar. In particular, the explicit introduction of non-bonded forces in the MMFF method, intuitively connected to the concept of "steric strain", is anything but standard technique among spectroscopists. A notable exception, which shows that the exclusion of non-bonded interactions may be more a matter of tradition than of physical unsoundness, is found in a recent study by Morino.[18] First, he assumes the usual potentials and uses anharmonic corrections only for the bonded atom pairs (which corrections are transferred from the corresponding diatomic molecule or radical). It turned out that the remaining anharmonic terms were quite well reproduced by introducing interactions between non-bonded atoms. The resulting "van der Waals-Valence Force model" brings the calculated values close to the observed ones, at least for small molecules. An even more fundamental approach was pioneered by Lifson and coworkers [19-21] in their search for a "self-consistent" force-field, which ideally should enable one to calculate relative energy, geometry and vibrational frequencies for a given class of compounds from a single set of parameters. We will return to this point below.

In the search for approximate potential functions the emphasis lies on the heuristic power of such functions in calculating selected properties of complex molecules. At this stage one is confronted with a host of problems and many quite arbitrary choices have to be made. These concern the exact form of the required parametrization and seems mainly guided by the goals and preferences of the investigator using his computer program and to a certain extent by the limitations inherent to the program and to the computer to be used.

Given a flexible computer program and a large memory-storage capacity machine, what can one do? In the ideal case one would like to use an internally consistent force-field which embodies *all* the following features:

(A) Correct prediction, *i.e.* calculation within, if possible, the limits of experimental error of a long list of properties:

    (a) *Properties pertaining to energy*
      — heats of formation, "strain" energies
      — conformational energies
      — heights of barriers separating various conformations, shape of potential energy wells and barriers separating conformations

    (b) *Properties pertaining to geometry*
      — geometry of molecules in their various conformations (bond distances, bond angles, torsion angles) in the ground state, *i.e.* in the hypothetical vibrationless state

     — geometry of molecules in the crystalline environment
     — geometry of molecules at the saddle point of the barrier separating conformers
     — geometry of molecules "somewhere" along the transition coordinate

(c) *Properties pertaining to molecular vibrations*
     — vibrational frequencies
     — amplitudes of vibration
     — torsional modes
     — thermodynamic functions (enthalpy and entropy of vibration)

(d) *Other properties that may depend on one or more of those above*
     — charge distribution and dipole moment, obtainable from the output geometry using semi-empirical quantum mechanical methods
     — nmr chemical shifts
     — chemical reactivity, relief of strain

(B) Coverage of all or most of organic structures with a reasonably limited set of transferable "force constants".

However, in hard practice it turns out that the realization of this lofty goal is yet hidden in the clouds, in other words, each investigator follows his personal preferences, chemical/physical intuition, and mathematical genius. Certain limitations have to be accepted. The present status is admirably summarized by Allinger *et al.*[22]: "*The mechanical model that we use to represent the molecule is at present very crude in comparison to the complex elegance actually dictated by the Schrödinger equation for the total wave function. We obviously cannot expect our present model to reproduce accurately all of the properties of a molecule, and have chosen to parametrize our model so as to fit (1) structure and (2) energy at 25 °C, and are prepared to sacrifice accuracy for other quantities to some extent as necessary. This approach is based on expediency. We need a reliable way to determine routinely structures and energies for use in other work*".

The approximation involved in factorization of the total wave function of a molecule into electronic, vibrational and rotational parts is known as the Born-Oppenheimer approximation. Furthermore, the Schrödinger equation for the vibrational wave function (which is the only part considered here), transformed to the normal coordinates Q_i (which are linear functions of the "*infinitesimal*" displacements q_i) yields equations of the harmonic oscillator type. For these reasons Lifson and Warshel [19] have stressed that the force-field calculations should not be considered as classical-mechanical in contradiction to quantum-mechanical, but rather as an inductive method, seeking a common analytical representation of a large set of observable data. The term *"molecular mechanics"* should be understood in this sense.

The sequence given above of properties pertaining to *energy, geometry* and *vibrations* follows from the basic assumption that only *small* amplitudes of motion are involved and that consequently the complete potential energy expression may be expanded in a Taylor series in powers of coordinates q_i which give the displacement of atoms from their equilibrium configuration. Thus, when all q_i's are zero the molecule in a given conformation is at its minimum potential energy:

$$V(q_1 \ldots q_i) = V_0 + \sum_i (\delta V/\delta q_i)_0 q_i + 1/2 \sum_i \sum_j b_{ij} q_i q_j + \cdots \cdots \tag{1}$$

with $b_{ij} = (\delta^2 V/\delta q_i \, \delta q_j)$.

In spectroscopic analysis the first term, V_0, is usually made zero by definition. This is done by assuming that *each* internal coordinate, let it be a given bond distance or bond angle, is "strainless" at its equilibrium value. For example, the "standard" sp^3-sp^3 carbon-carbon bond length r_0 is 1.53 Å. The harmonic stretching potential then is represented by:

$$V_r = 1/2 \, K_r \, (r_i - r_0)^2 \quad \text{with } r_0 = 1.53. \tag{2}$$

In other words, in normal coordinate analysis r_0 is not regarded as an adjustable parameter but set equal to the value assigned to the given bond in the molecule the spectrum of which is to be calculated. Similarly, in the corresponding expression for bond angle deformation of, say, a C—C—C bond in a hydrocarbon, θ_0 is set equal to the tetrahedral angle. Non-bonded repulsions between geminal atoms are either accounted for by a harmonic potential (Urey-Bradley or UBFF) or by introducing General Valence Force Field (GVFF) interaction constants expressed in terms of internal coordinates (f_{rr}, f_{ra}). Repulsions between atoms not bound to a common atom are simply ignored, which means that their effect is absorbed by the remaining parameters. For these reasons "spectroscopic" UBFF or GVFF constants are in general incompatible with the fields used in molecular mechanics calculations (MMFF), as was pointed out by several authors. [19,23-25] This fact is especially important in crowded or otherwise strained molecules where non-bonded atom-atom repulsions cause bonds to stretch and angles to bend.[24] Therefore, in a molecule of suitable complexity one may encounter "long" and "short" carbon-carbon bonds, long bonds being indicative of local steric strain, and widely varying values for the nominally "tetrahedral" bond angles. Several authors independently have come to the conclusion that one of the main conditions to be imposed on a "good" MMFF, that of complete transferability of the set of parameters to the greatest possible number of molecular systems, strained and (relatively) unstrained, necessitates a judicious choice of the so-called "natural" or "strainless" values of r_0 and θ_0 in conjunction with the determination of k_r and k_θ and with the chosen form of the FF. In case one wishes to use the classical UB expression to account for the van der Waals interaction between geminal atoms:

$$V_R = 1/2\, F_{ij}\, (R_{ij} - R_0)^2 \qquad (3)$$

the "equilibrium" distance R_0 also must be seen as an adaptable parameter.

More will be said about this point in the chapter devoted to functions and parameters, let us return for the moment to Eq. (1) and the physical meaning attributed to each of its terms. In order to facilitate the discussion we shall cast Eq. (1) in the form:

$$V(q_i) = V_0 + \sum_i F_i'\, q_i + 1/2 \sum_i \sum_j F_{ij}''\, q_i\, q_j \ldots\ldots \qquad (4)$$

The equilibrium configuration is mathematically defined as the set of q_i values for which $V(q_i)$ is a minimum and *physically defined as the configuration at which the total spring forces* $\sum F_i'$ *(first derivatives) balance off exactly for each particle*, that is, the second term in (4) is zero in the equilibrium configuration, the so-called hypothetical motionless state.[26] The matrix of second derivatives (third term) then yields the vibrational frequencies.

In molecular dynamics we have no *a priori* knowledge of the spring forces F_i', we must gain such knowledge from available experimental information. Now it is clear that we require three types of information simultaneously to ensure proper scaling and balance; furthermore, *we require this information for families of molecules:*

(i) *Energies*, or rather experimental energy differences between conformations, geometrical isomers and heats of formation as a check upon the calculated values of V_0. It should be added that V_0 represents to a certain extent a *measure* of the strain energy, but should not be equated to it. A proper description should be "Force Field Strain Energy", being dependent on the chosen functions and parameters. V_0 can be manipulated up to a certain extent without changing the balance of forces (= geometry) or the vibration frequencies.

(ii) *Experimental geometries*, especially of molecules in which a certain kind of strain is expected to have predominance. Statements sometimes encountered to the effect that the information content of a particular geometry is small need some further qualification. It is true that the geometry of simple alifatics, for example, or of alicyclics, can be reproduced almost exactly by widely different sets of force field parameters. In these cases any reasonable set, however simple, seems to work. In fact, success is already guaranteed by setting the "natural" bond lengths, -angles and other null-parameters (Eqs. 2 and 3) equal to the experimental values, in conjunction with a choice of sufficiently soft non-bonded potentials and the use of spectroscopic force constants. However, such a simple field is bound to fail sooner or later for molecules that are either highly crowed or are highly deformed because of conflicting demands imposed, e.g. for certain bicyclic systems. In our opinion, no proposed set of force field parameters may lay claim to more general validity before it has been shown to yield good geometries in severely strained as well as unstrained systems. Thus seen, the information content of geometries of strained molecules is of crucial importance, be it in a negative manner, in the course of the process of selecting proper potential functions and transferable parameters. Unfortunately, systematic and accurate experimental investigations in the area (viz. X-ray and electron diffraction studies of highly strained hydrocarbons) are as yet relatively rare. Notable exceptions, among others, can be found in the work of Bartell [27] and of Dunitz et al.[28-30]. We note that is has been stated recently [31] that even the exact geometry of a key bicyclic compound, norbornane, has not been experimentally established beyond doubt. Further illustrations of these points will be given below in our chapter on applications of force-fields methods to structural problems.

(iii) *Vibrational frequencies.* Although it should be clear from the foregoing that knowledge of vibrational frequencies of molecules constitutes the third leg upon which a reliable and chemically perspicuous force field should be build, thus far only two groups of workers, Lifson et al. [19-21], and Boyd et al. [26,32], have simultaneously and independently, followed this admittedly difficult route. These authors emphasize that valence functions used should lead to results that are simultaneously *consistent* insofar as possible with the three general properties mentioned: relative energy (heats of formation), geometry and vibrational frequencies. This overall view is regarded vital even at the *sacrifice of some detailed agreement* in any one of these areas alone.[26] This approach has been denoted CFF (consistent force field [19-21] method), the term "consistent" emphasizing the main feature of this method, namely the simultaneous calculation of different properties of families of molecules. The information content of a vibration spectrum by itself, as source of inquiry into the exact form of the potential functions governing the dynamics of a molecule of some complexity, is probably not very great for several reasons. In fact, a statement has appeared [21] to the effect that requirements of the conformational and vibrational properties are contradictory.

The problem of multiple force constant solutions arises directly from the ambiguities often encountered when assigning the observed frequencies. Furthermore, the earliest CFF calculations [19] already showed that the vibration spectra of a family of cycloalkanes could be reasonably well reproduced by different sets of force constants within the framework of a given assignment (the parameter sets being calculated using slightly different functional forms of the potential energy expressions).

In principle, calculated frequencies can be manipulated to some extent without significantly affecting calculated geometries by the employment of certain potential functions (*e. g.* the term describing the interaction of angle bending and torsional rotation [21]) that combine the properties of having a fairly large value of the second derivative and a small or negligible value of the first derivative. In our opinion any physically sensible manipulation of this kind is legal as long as criteria of transferability of force constants and reproducibility of energy and geometry are met.

It should be stressed that adherence to this overall view is not strictly necessary when it comes to computerized problem-solving in organic chemistry. Indeed, most authors prefer to follow the simpler course set by the pioneering work of Hendrickson [10,33–38] and Wiberg [39] and pursued in depth by Allinger *et al.* [22,40,41], a course essentially inspired by Westheimers approach.[42,43] In the Hendrickson-Wiberg-Allinger (HWA) method the "force field" *energy* Vq_i is minimized with respect to each internal coordinate in an iteration process, which process, in its latest form, is reported [22] to converge faster than does the matrix method (based on finding the balance between the spring forces) embodied in Boyd's program. [26,32,44] Of course, spectral information cannot be utilized in the HWA method, nor can spectral data be predicted. As a consequence, the HWA force-field parameters are scaled slightly differently from those obtained in the CFF method. In the latter, the calculations are carried out to obtain the hypothetical motionless state, from there, it is a simple matter (provided the calculated spectra are good enough) to add the temperature-dependent vibration-rotation-translation terms (V_{vrt}) [19,32,44] in addition to temperature-independent group-contributions in order to predict properties like the heat of formation or conformational energies at any desired temperature. In the HWA method parametrization is aimed at reproducing experimental heats of formation among other properties at a selected temperature, usually 25 °C, hence V_{vrt} is implicitly absorbed in either or both the calculated V_0 or in the group contributions. Fortunately, V_{vrt} of many hydrocarbon molecules may be approximated with reasonable accuracy by taking its contribution as an additive property for which group increments can be worked out.[26] On doing this one has to take care to include cyclic as well as non-cyclic model structures in the basic set that is used to derive force field parameters and group contributions, because it has been found [26] that enthalpy and zero point energy group increments from a non-cyclic series tend to overestimate the values for cyclic molecules. This finding partially explains the success of the HWA approach. Of course, the sum of the group contributions is the same for a set of conformational isomers and, if the additivity rule holds at all, one may expect that the contribution of the V_{vrt} term to conformational energies is relatively small. This actually seems to be case [21], at least for simple systems like n-butane and methylcyclohexane (< 0.1 kcal/mole). It remains to be seen whether or not neglect of V_{vrt} is permitted in larger or more highly strained molecules. It should be mentioned that as yet computer limitations prohibit calculation of the vibrational contribution in molecules containing over 40 atoms or so, but that there is no reason to neglect the rotational contribution for any system, whatever its size, because its calculation is simple and straightforward. Besides, an interesting spin-off is the calculation of the principal moments of inertia and the possibility of carrying out automatically the transformation of the (usually arbitrary) cartesian coordinates of the molecule onto the unique principal axes system. The moments of inertia product is a number that is experimentally accessible from microwave spectroscopy, even if complete analysis of the molecular structure is out of the question, and may be compared directly with the calculated value. Another (future) advantage of this transformation to principal axes may lie in uniform comparison of the results of different force fields and in uniform presentation of coordinates for storage purposes, *e.g.* in a central data bank or on microfilm.

It has been said above that in molecular mechanics V_0 represents a certain measure of the "strain energy" of the molecule in the hypothetical motionless state. It should be realized that V_0 for a given molecule is a *function of the FF-parametrization* and does not represent by itself a physically measurable quantity [35], but of course differences in calculated V_0 values of conformers or geometrical isomers are experimentally verifiable and constitute a relative energy scale, supplemented by group contributions of some sort or another to build an absolute scale from which heats of formation are predicted. This means that the group contributions themselves are force field dependent. At first sight one may suppose that V_0 represents a physically meaningful sum of deformation (stretching, bending and twisting) energies plus the non-bonded energies and that independent physical meaning may be ascribed to the *individual* energy functions in a given well balanced force field. Up to a certain extent this may be true, however, lacking an objective yardstick by which one can measure a *single potential* in a complex molecule, *in the absence of other potentials*, it is hard to see on what basis one is justified to discuss quantitatively "origins of strain" in a given molecular species. Here we touch upon an important question that merits some thought. Let us give an example of the problems encountered. The calculation of the medium rings C8—C10 has long been in the center of interest. These rings are characterized by "abnormal" values of bond angles, fairly long carbon-carbon bonds, torsion angles far removed from the staggered minimum energy configuration and last but not least by extremely short H...H nonbonded distances. The correct calculation of the energy, geometry and vibration spectrum of these systems constitutes an interesting (but not exclusive, nor conclusive) challenge for the "quality" of any proposed force field. Several authors have carried out these calculations and discussed the partitioning of strain in terms of quantitatively distinct contributions of stretch, bend, torsion and van der Waals energies. We take the view that such partitioning cannot be ascribed exact physical meaning except in a *grosso modo* sense. Different force fields yield different partitions for the simple reason that underestimations of certain parameters are compensated for by overestimations elsewhere. All one can do is placing upper and lower limits on each individual contribution in the absence of independent information. Certainly, angle strain is present in cyclodecane for example, more than in cyclohexane, but how much? An upper limit is evidently set by the total experimental strain energy, minus perhaps the reasonably well known torsion strain, the lower limit is (close to) zero. A similar argument applies to the energy absorbed by the stretching of the various bonds and by the pressing together of hydrogen atoms. The logical conclusion must be that the "hardest" function or functions adopted (as far as energy is concerned) will tend to absorb most of the strain. Because of the strong anharmonicity of the non-bonded potentials this need not be strictly true for the strain "forces" (first derivatives) acting on the geometry. Seen in this light the current controversy concerning the "origin of strain" in the adamantane molecule loses much of its meaning. Schleyer et al. [45], using spectroscopic force constants (stretch, bend, torsion) added a "best set" of van der Waals H...H, C...H and C...C potentials needed in order to arrive at the experimental strain energy of adamantane (about 5—6 kcal/mole), and finally concluded that this procedure, besides yielding a good direct measure of C...C function hardness [17,45], indicates that C...C interactions play a major role in the built-up of strain in this cage molecule. This procedure, for one thing, lacks self-consistency in that addition of non-bonded potentials to a set of spectroscopic force constants in fact renders the latter set invalid as far as calculation of the vibrational frequencies is concerned [26]. Furthermore, only three true non-bonded (1,5) C...C interactions occur in adamantane, the remainder being of the 1,4 type. Now 1,4 interactions pose a special problem because *ab initio* calculations have shown that at least part of the torsion barriers is due to pair-pair interaction between electrons (*bonds*) over and above van der Waals interactions. Therefore, the validity of using adamantane as a measure of non-bonded interactions remains doubtful. Allinger et al. [22], on the other hand, selected a set of non-bonded potentials from independent sources

13

(properties of a hydrocarbon crystal), and supplemented their set with stretch, bend and other parameters to arrive at a field that produces good heats of formation and reasonable geometries (not spectra) of hydrocarbons in general. This force field, used to calculate adamantane, led the authors to the conclusion that the strain in this molecule is due to an excessive number of H...H repulsions and a decreased number of C...H and C...C repulsions, just as in cyclohexane.

The present authors feel that debates of this kind rest on the several unwarranted assumptions mentioned above. If one endeavors to assess published model force fields by carrying out calculations on various types of hydrocarbons, employing different fields, one is struck by the fact, already commented upon in a slightly different context by Burgi and Bartell [27], that analyses to find which of the interactions bear the main burden of stress depends on the force field assumed to be valid. For the time being one should avoid attempts to obtain more information from the model calculations than is really warranted. Another important case in point is the theoretical analysis of tri-*tert*-butyl-methane, a molecule with exceptional crowding.[46]

Is there a way out of this dilemma? We are inclined to answer this question in a positive sense. First of all, recourse to theory is called for. It has been stated [17] that the very idea of a strain-free or preferred structural parameter is ill conceived, but viewing the problem from a dynamic standpoint one sees that one is really concerned with finding the balance between spring forces. Even in simple molecules like methane such a balance, *i.e* strain, exists (a fact already recognized by Simanouti [47] in 1949 and called by him "intramolecular tension") and closer analysis of these balances, coupled with *ab initio* calculations of non-optimal geometries might provide much-needed insight. One should keep in mind that the undeniable successes of the MMFF method obtained thus far (based on the judicious choice of simple reference molecules) show that the basic premise of the method, the possibility to extrapolate from simple to more complex molecules and obtain reliable information at low cost, holds sufficiently well to warrant further exploration.

Secondly, critical evaluation of published force fields by carrying out comparative calculations on admittedly "difficult" systems may indicate weak points, *i.e.* at least which functions are too "hard" or too "soft" on a relative scale. Of interest in this respect are systems containing regions of severe crowding as well as systems on which conflicting demands are imposed, *e.g.* bicyclics. A serious problem often encountered is the lack of really accurate experimental information for molecules containing features that would provide a crucial test of the validity of a *specific* potential function. A more general testing ground is now available in the geometry of steroids, for which class of compounds so much experimental (X-ray) information is at hand that is has been possible [48] to draw an accurate picture of the "standard" hydrocarbon skeleton, *i.e.* the weighted average of over 60 individual determinations. Comparison of this standard steroid geometry with results obtained with various force fields is given below.

A remark on the question as to where one should draw the borderline separating systems within the range of a given parametrization and those outside this range is still in order. Important cases in point are the smaller rings. Allinger *et al.* [22,41] have consistently taken the view that cyclobutane should be included in the set of structural types over which the force field should be optimized. Their "natural" C—C—C bond angle for methylene groups (110.2 °C) is therefore also used in the calculation of four-membered rings. The increasing carbon-carbon bond distance with decreasing bond angle is taken care of by introducing a stretch-bend interaction. [22,25] Molecules more distorted than cyclobutane are considered separately. Boyd *et al.* [26] have considered the cyclobutane and cyclopropane rings to be special electronic systems with their own specific force constants. Accordingly, Boyd *et al.* take the equilibrium C—C—C angle in cyclobutane to be 90 °C. However, an ambiguity then results, when building multicyclic systems, in selecting "natural angles" at the juncture atoms when a $n = 3$ or $n = 4$ ring is fused to a larger ring. The theoretical ramifications of this problem have not been worked out, but seem to be of great importance.

III. Functions and Parametrization

As early as 1960 Bartell [49] stated: *"it will be interesting to see if simple potential functions will be forthcoming which are capable of correlating quantitatively bond angles and spectroscopic results as well as thermochemical results"*. Indeed, many functions came forth in recent years that proved extremely useful but much work remains to be done, even the hydrocarbon problem has not yet been solved optimally. One of the underlying reasons for the the relative lack of progress seems to be that sufficiently diverse and accurate experimental data, necessary to generate the model force field, simply do not exist. The problem of accuracy is crucial. For example, it has little sense nowadays to compare a calculated geometry of a complex molecule, say of a steroid, with an X-ray determination which has not been carried out to a higher accuracy then $\sigma = 0.01$ Å (nominal standard deviation) in the carbon-carbon bond lengths. This is so because the usual rule of thumb states that the differences should be of the order of three times the standard deviation before they may be regarded as significant, and any recent force field allows prediction of C—C bond lengths at the 0.01—0.02 Å accuracy level or better. Similar remarks pertain to bond angles and torsional angles, and to thermochemical data. We have pointed out repeatedly that each of the three general properties, energy, geometry and frequencies can be manipulated by itself up to a certain extent without changing the results obtained for for the other two. For instance, the contribution of a given deformation (q_i-q_0) to the deformation energy V may be modified in several ways without changing the corresponding geometry and vibrational frequencies and vice versa. One such method is to introduce a cubic term [22] (of special value for large deformations of bond angles) [17], another is to introduce a linear force constant. The latter practice is standard in normal coordinate analyses when Urey-Bradley 1,3-potentials are employed.

Let us suppose that a given geometry parameter q_i requires, within the framework of a certain force field, a given force $(F_q)_i$ in order to balance off correctly (Fig. 1). It is easy to see that an infinite number of combinations

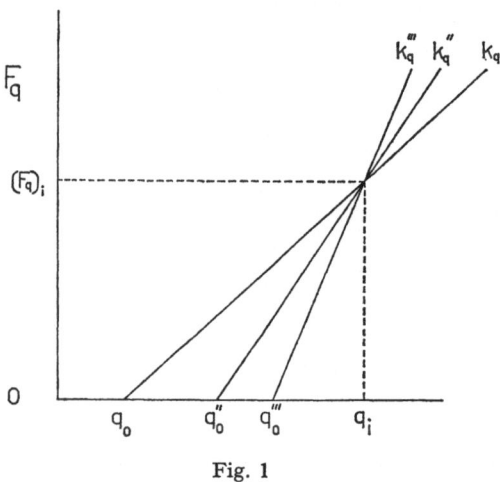

Fig. 1

of k_q and q_0 will do the job, the only requirement being that the force at point q_i is equal to $(F_q)_i$. There are other consequences however. The choice (k_q, q_0) implies a relatively "soft" function, the choice (k_q''',q_0''') a relatively "hard" one. The associated vibrational frequency, keeping all other FF parameters constant, will depend on our choice of k_q. On first sight one would be tempted to take k_q (rather, the total F-matrix) from measured frequencies and then adapt q_0 to get a good match with observed geometry. This method works quite well for relatively unstrained hydrocarbon molecules where bond distances and angles remain close to the optimum values. One runs into difficulties with strained systems and here linear terms have been useful in practice, because such a term allows one to retain the desired softness or hardness properties (and the desired frequencies) and still match the calculated energy with the observed one.

(i) We will next concentrate our discussion on bond stretching and -compression, especially on the carbon-carbon bond for reasons of convenience, since a wealth of experimental and theoretical information on the variation of the C—C bond length with chemical surroundings is available.

In the earliest work in the field of molecular mechanics (Hendrickson [10]) bond distances were not allowed to vary, the C—C bond was kept at 1.533 Å. The assumption of invariant C—C bonds was based on the known fact that bond bending is energetically cheaper than bond stretching. Thus a molecule under moderate strain will first strive to adapt its bond angles, not much relief of strain can be gained by adaptation of C—C and C—H bond lengths. Wiberg [39] seems to have been the first author to incorporate bond lengths distortion in an energy minimization scheme, employing a rather "stiff" harmonic potential:

$$\Delta V = 1/2 \, k_r \, (r - r_0)^2 = 1/2 \, k_r \, \Delta r^2 \qquad (5)$$

where r_0 is the "equilibrium" bond length (1.54 Å for C—C bonds) and k_r the stretching force constant evaluated from vibrational spectral data (about 5 mdyn/Å^2).

The "standard" (1.54 Å) carbon-carbon bond length at the time (1965) of this work was of course well reproduced by Eq. (5). Boyd [44] similarly adopted a stiff potential ($k_r = 4.6$ mdyn/Å^2, $r_0 = 1.54$ Å) from Snyder and Schachtschneider's spectroscopic work on hydrocarbons [50]. Allinger et al. [40,41] introduced an important refinement in regarding r_0 (CC) to be an *adjustable parameter* ("natural" length)[a], chosen so as to accurately reproduce the C—C bond distances in a series of relatively unstrained saturated hydrocarbons. All bond lengths were calculated [41] within 0.01 Å of the experimental microwave values (r_s) by taking r_0(CC) = 1.513 Å (for primary and secondary carbons) or 1.509 Å (for tertiary and quaternary carbon on either end), and retaining the spectroscopic k_r given above. In this review each carbon-carbon bond length is the resultant of the non-bonded (1,4) repulsions between the atoms or groups situated at each end of the C—C bond in question pushing the carbons apart and the carbon-carbon bond trying to pull the two carbon atoms back toward the "natural" distance. Thus, even the C—C bond in ethane is "strained" to some extent (the notion that "non-bonded" interaction were mainly responsible for observed variations in C—C bond lengths and -angles was strongly advocated by Professor L. S. Bartell during the 6th International Congress of the International Union of Crystallography, September 1963, but this idea met very little favor at the time [51]). In the actual example of the force field

---

[a] it is of interest to note the various verbal descriptions of r_0: "natural" length [26,41], "strainless" length [22], "optimum" length [36], "reference" length [24].

employed (AFF3 [41]) the C—C bond in ethane is stretched from 1.513 to 1.527 Å by the H...H 1,4 repulsions. On eclipsing, the hydrogen atoms necessarily come closer together, resulting in opening of the C—C—H bond angles as well as in predicted stretching of the central carbon-carbon bond (to 1.533 Å), *i.e.* an increase of 0.006 Å. It is of interest to note that MINDO/2 calculations [14], where the energy of the staggered and eclipsed forms of ethane was minimized with respect to all other coordinates, a similar trend was reported, although the absolute values are slightly smaller than the experimental ones (C—C staggered 1.487 Å, C—C eclipsed 1.494 Å). However, in subsequent work by various authors important deviations from this simple scheme were devised in order to account for its deficiencies.

First, it was noted by Jacob, Thomson and Bartell [24] (JTB) that it is inconsistent to use spectroscopic valence force constants in a field that includes all pairwise non-bonded interactions. They proceed by assuming that the forces governing a particular kind of internal coordinate, *e.g.* a C—C bond, are identical in all hydrocarbon molecules, except for its non-bonded environment. It should be mentioned in passing that the JTB field includes certain strong Urey-Bradley (UB or 1,3 interactions, *i.e.* interactions between atoms bound to a common atom) whereas in the older fields all important UB interactions were purposely neglected. However, the JCB-UB potentials ("ordinary" non-bonded interactions) are not completely consistent with the empirical UB force constants of Snyder and Schacht-schneider. [50] JTB suggest that perhaps the least arbitrary way to minimize the discrepancy would be to *rederive* a set of stretch and bend constants which, together with the non-bonded potentials adopted, would best fit the molecular spectra (more recently, Burgi and Bartell [27,46] argued that bond anharmonicity, which has only a minor effect in structure calculations, plays a vital role in frequency calculations). Their "reference" r_0 (CC) value, based on the structures of ethane and methane, came out surprisingly small (1.24 Å), but combined with a k_r(CC) of 2.2 mdyn/Å, the total field accounted reasonably well for observed variation in C—C bond lengths in a series of hindered open-chain alkanes. [24] No attempt was made to faithfully reproduce conformational energies, heats of formation, or vibrational frequencies.

The recent inclusion of a specific stretch-bend interaction in calculations of molecular geometry and energy by Allinger *et al.* [22] did not necessitate major revisions as far as the stretching constants or r_0 are concerned. It is known from vibrational analyses of alkanes and cycloalkanes that when force fields specifically include Urey-Bradley forces, this must be accompanied by a vast reduction of the C—C stretching constant in order to reproduce the experimental C—C stretching frequencies. However, reduction of k_{CC} also necessitates a substantial reduction of r_0(CC) in calculation of geometry because the balance between the sum of the non-bonded and Urey-Bradley forces on the one hand and the force that opposes stretching of the carbon-carbon bond on the other should faithfully reproduce known variations in C—C bond lengths.

It is known that C—C bonds show considerable stretching when the C—C—C bond angles are forced to decrease from the approximately tetrahedral value, but whether or not these bonds actually should stretch on *increasing* the valence angle remains a matter of debate. The bond length in cyclohexane is 1.528 Å ($\theta_{av} = 111$ °) and 1.546 Å for cyclopentane ($\theta_{av} = 104.4$ °, $\varphi_{av} = 26.5$ °). [52] On the other hand, the average C—C bond length in various cyclooctanes is only slightly greater than the cyclohexane value (1.532 Å, $\theta_{av} = 116.5$ °, $\varphi_{av} = 43$ °) [53] and this increase may be due to various causes beside bond angle bending, such as 1,4 and other non-bonded interactions. Thus, it is at least premature to assume a stretch-bend interaction which has the effect of stretching the bond regardless of angle opening or angle closing [22]. The simple harmonic UB interaction on the contrary has the form:

$$V_{UB} = 1/2\, F_{ij}\, (R_{ij} - R_0)^2 \qquad (6)$$

and in the calculation of structures has the effect of introducing forces directed along the C—C bond. These forces are directed such that the bond stretches on angle closing and contracts on angle opening. However, no clearcut example of this behaviour at large C—C—C bond angles is known to us. A way out of this dilemma might be found along the lines proposed by JTB, *i.e.* to assume a non-harmonic[31] 1,3-interaction of the Lennard-Jones 6/12 Buckingham exponential type.

A problem concerning the carbon-hydrogen bond has been described. [21,22] Since the electron density in these cases is not centered at the H-nucleus itself (some of it being shifted inward into the overlap region), it is not correct to use the coordinates of the nucleus in calculations of van der Waals interactions. Williams [54,55] suggested that an offset of about 10% (in the direction of the atom to which the hydrogen is bonded) is necessary to explain certain crystal packing phenomena. Therefore it seems best to use the center of electron density to calculate van der Waals interactions and the position of the hydrogen nucleus for other properties. [21,22, 54,55]

(ii) *Internal Rotation.* The energy of a molecule changes with rotation of atoms or groups of atoms about an interconnecting bond. In the case of formally single bonds the activation energy necessary to pass the top of the barrier opposing free rotation usually amounts only a few kcal/mole. At first sight the insertion of potential curves describing internal rotation in MMFF calculations seems simple and straightforward, but a closer look reveals the existence of certain problems which need some consideration.

Let us first examine current practice in nomenclature of the phenom-menon of internal rotation. Standard textbooks on conformational analysis on the one hand and those on spectroscopy on the other hand deal differ-ently with the basic definitions. Internal rotation is measured by one or more torsion angles (dihedral angles, azimuthal angles). In the simple cases of *e.g.* a 1,2-disubstituted ethane the potential energy associated with the internal rotation may be written in a formal sense as a truncated Fourier expansion:

$$V_\phi = 1/2 \sum_{n=1}^{N} V_n^\circ (1 + \cos|n\phi|). \tag{7}$$

The signs of $\cos(n\phi)$ and of the V_n components may be chosen in various ways. Spectroscopists seem to prefer defining the zero angle of rotation to correspond to the staggered *anti* rotamer, but in conformational analysis one always defines the zero angle of any one torsion interaction as the eclipsed situation. The signs of the angles are taken *positive* when the rear atom has rotated away in a clockwise sense with respect to the front atom, as shown below:

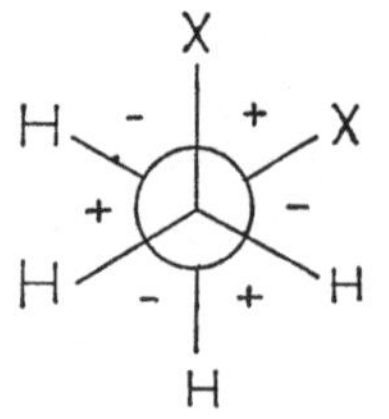

Early in the history of MMFF calculations [10,33,34] it was realized that the potential function of internal rotation may be considered as the sum of two contributions:

a) The non-bonded interaction between vicinal atoms
b) The "intrinsic" torsion interaction.

In practical evaluation of force-fields parameters the H...H and C...C non-bonded potentials were chosen first. Calculations of staggered and eclipsed rotamers of ethane, propane and n-butane then revealed the contribution of the non-bonded interactions to the total rotation barrier. For example, the van der Waals parameters proposed by Allinger *et al.* in their first paper [40] on calculation of conformations were such that the repul-sion between hydrogens on adjacent carbon atoms accounted for about 31% of the barrier in ethane. The remainder was accounted for as a quantity which was added by considering the torsion interaction to be zero for all angles of 60° or greater, and letting the energy increase with decreasing ϕ according to

$$V_T = 1/2 \, V_3^{\circ} \, (1 + \cos 3\phi) \tag{8}$$

with V^0 selected to be one third of the intrinsic torsion energy for each of the eclipsing pairs of hydrogens. A similar rule of additivity of non-bonded and intrinsic torsion energies has been shown by other workers [21,26] to yield reasonable vibrational frequencies, although a notably better agreement with experiment was obtained by adding a bond angle-torsion angle cross term [21]. The presence of hetero atoms or double bonds in the molecule to be calculated poses several special problems. One may have to introduce attractive terms into the Fourier expansion; $V1$ and $V2$ terms must be scaled somehow (scaling is important as soon as barriers no longer consist of repulsions in the MMFF sense) and finally, out-of-plane bending potentials in addition to pure torsion must be added to ensure proper geometry of ethylenic compounds and ketones.

A more fundamental dilemma is concerned with the physical significance of the approach described, *i.e.* whether one should calculate barriers on the basis of complete relaxation or on the basis of a "rigid rotor" approach. [22] In the first case the calculated barrier is simply taken as the energy difference between the eclipsed and staggered conformations, allowing the molecule to relax in all degrees of freedom in each form. In other words, complete coupling between the torsional mode and stretching and bending modes is assumed. The other extreme assumption, that of the rigid rotor, implies that bond distances and bond angles do not change during rotation. Allinger *et al.* [22] propose a way out of this difficulty by calculating the barriers on the basis of complete relaxation but purposely letting them come out *lower* than measured.

At least part of the trouble stems from the fact that up till now no clear distinction has been made between the *dynamic increase* in energy of a molecule occupying a high vibration level in the torsional potential well (*e.g.* approximately at the top of the barrier) with its associated large amplitude of motion on the one hand and the *static* energy increase of part of a molecule *constrained* to be eclipsed on the other hand, *e.g.* as in bicyclo[2.2.1]heptane, both compared to the energy content of the staggered geometry. It seems more correct to study the static (constrained) cases separately. The rigid rotor model gives the *maximum* value of the barrier, the experimental value should be lower, the complete relaxation model perhaps represents the *minimum* value.

In the absence of a mathematical model that can take care of the large amplitudes of motion encountered in the higher torsional levels (note that the power expansion on which the MMFF method is based presupposes infinitely *small* amplitudes) the calculation of the exact geometry and energy of *e.g.*

ethane in the hypothetical motionless *eclipsed* state may be more of an exercise than physically meaningful. It seems that these difficulties can be avoided in part by setting the V_n° values equal to the observed (spectroscopic) barriers, thus omitting the unreliable contribution of non-bonded interactions to the total torsion strain. Such a procedure of course necessitates a judicious separation of contributions of H/H, C/C and C/H angles as well as introduction of a stretch-torsion cross term.

(iii) *Bond Angle Bending.* The treatment of bond angle deformation in MMFF calculations has been quite diversified in the past. It was soon apparent that the bland use of spectroscopic force constants and tetrahedral "natural" angles in a harmonic approximation could lead to grossly overestimated strain energies at large angle deformations. Several solutions to this problem have been proposed. Hybridization arguments [17,22,40] were invoked in order to justify the use of different "natural" angles, assumed or experimental, for various substitution types of the central carbon atom, or a linear bending term assumed [21], which has essentially the same effect. A sigmoid curve with empirical parameters has also been used [41], but this particular curve suffered from the defect that its first derivative changed sharply at specific values of $\Delta\theta$. The trend now is to use rather smaller values of the bending constants than previously, coupled to the introduction of a cubic term which reduces the effective k_θ at large deformations [17,22].

Intimately connected to the choice of bending parameters is the choice between a general valence force (GVFF) expression (including a stretch-bend interaction) or a UBFF representation. The necessity for the inclusion of either a stretch-bend or a Urey-Bradley term to account for the abnormally long bonds in *e.g.* norbornane was pointed out by several authors [17,22,31]. Several representations have been proposed: "normal" non-bonded interactions [24], the classical UB expression [21] and whittled atoms [7,31,56] (smaller radii in the direction of the geminal atoms), but a further theoretical analysis of the UB potentials seems mandatory.

(iiii) *Non-Bonded Potentials.* This subject has been treated exhaustively in Ref. [17] and we shall be content to discuss the subject briefly. In the MMFF approach a pairwise additivity of repulsive and attractive interactions between spherical atoms is assumed. From the foregoing it is clear that no consensus of opinion exists concerning the inclusion of non-bonded potentials in 1,3 and 1,4 interactions. On the other hand the evidence of strain in sterically crowded molecules necessitates the adoption of some sort of repulsion between atoms. Perhaps unfortunately, no clear guidelines have appeared on which to base a more or less reliable choice of the functions and parameters to be used. Most workers have assumed that interatomic and intermolecular (isotropic) potentials provide a quantitative model for intramolecular potentials, but this assumption has been strongly critisized [17] on grounds that the effects of the chemical environment between atoms is neglected (*i.e.* the problem of shielding of atoms from each other by interfering charge densities). On the other hand, potentials taken from crystal fields of hydrocarbons may be expected to include long distance interaction inclusive of "average" shielding. It may turn out that short distance interaction in situations of buttressing hydrogen atoms are perhaps best taken from 2nd virial coefficients. [57] Warshel and Lifson [19-21] use Coulombic terms between charges localized on the interacting atoms (in addition to the conventional non-bonded expressions) as adjustable parameters. We would definitely prefer to take charges, if necessary, from quantummechanical calculations. From our experience, charge-charge interactions hardly affect the calculated geometry, but may be used to manipulate the calculated V_0.

The functions most often used in MMFF calculations are either a two-parameter Lennard-Jones type:

$$V_{nb} = A / R^n - B / R^6 \qquad (9)$$

or take a three-parameter Buckingham exponential form:

$$V_{nb} = A \exp(-B R) - C / R^6. \qquad (10)$$

For practical use several authors [22,26,35] prefer the expressions due to Hill [58], which have been devised in such a fashion that only two parameters are necessary, r^* (a "radius") and ε (a measure of "softness" or "hardness"). Eqs. (9) and (10) then (for $n = 12$) yield essentially the same curve, except in the extreme repulsive part:

$$V_{nb} = \varepsilon (2r^*/R)^n - 2.0 \, \varepsilon (2r^*/R)^6 \qquad (11)$$

$$V_{nb} = 8.28 \times 10^5 \, \varepsilon \exp(-R/0.0736 \times 2r^*) - 2.25 \, \varepsilon (2r^*/R)^6. \qquad (12)$$

The minimum of these curves occur at $R/r^* = 1.0076$ with $V_{min} = -1.210 \, \varepsilon$, the energy passes through zero at $R/r^* = 0.894$. Overlooking the literature on hydrocarbon crystal fields one gains the strong impression that one of the main problems encountered is that of correctly separating the contributions of C...C, H...H and C...H to the total non-bonded potential. [54,55] The observational equations for non-aromatic crystal structures alone were insufficient to define these potentials, the combined non-aromatic and aromatic data [55] yielded reasonably looking potentials (incorporated in a MM force field developed by Boyd *et al.* [26]). The C...H interaction is usually fitted by a combining law from the C...C and H...H parameters, but Williams [55] produced strong evidence that this procedure should not be used for the all-important repulsion part of the non-bonded C...H curve.

In a later study, based on non-aromatic data only, Warshel and Lifson [21] reported that the attractive part of the Lennard-Jones potential is better represented by a $1/R^9$ rather than by a $1/R^{12}$ function.

In summary, much progress has been made during the past decade, but a general solution to the hydrocarbon problem has not been found. Gradually, however, a certain convergence of proposed parameters and functions to a a common denominator seems developing.

IV. Minimization Methods

In all methods the total strain energy V_{tot} is written as the sum of several types of energy contributions. In a general form:

$$V_{tot} = V_r(r) + V_\theta(\theta) + V_\phi(\phi) + V_\delta(\delta) + V_{ub}(d) + V_{nb}(d) + \\ + V_{coul}(d) + \text{crossterms} \tag{13}$$

where V_r, V_θ, V_ϕ and V_δ represent the total energies of bond length, bond angle, torsional and out-of-plane deformation respectively and V_{ub}, V_{nb} and V_{coul} stand for Urey-Bradley (1,3), nonbonded and Coulomb interaction energy contributions.

In order to simplify the problem, several workers have constructed force fields that include only some of the terms in Eq. (13). Usually crossterms are ignored although it should be stressed that they can be included in a straightforward way, e.g. the interaction term $V = k(\theta - \theta_0)(\theta' - \theta'_0)$ cos ϕ used by Warshel and Lifson [21], or the stretchbend interaction advocated by Allinger et al. [22].

Two different kinds of force fields have appeared in the literature. In the first type a set of internal coordinates, sufficient to describe the molecular geometry, is choosen. The total "strain" energy is then calculated from the deviation of these internal coordinates (plus nonbonded interactions) from their "strain free" values. The redundant coordinates are given zero force constants and do not contribute to the potential energy.

The second approach is less tied to spectroscopic habits; here each structural parameter is considered to contribute to the total energy. It shall be clear that the force constants used in both methods are not directly comparable without a suitable transformation.

Assuming that a set of parameters and functions is obtained, one next has to choose a "trial model" in terms of Cartesian or internal coordinates.

The structural parameters of this model (internal coordinates plus 1,3 distances plus other nonbonded distances) can now evaluated, directly from goniometric relations if Cartesian coordinates were used, or via a preliminary transformation to Cartesians if the input was given in internal coordinates. From Eq. (13) the "strain" energy of the trial model can be evaluated and this energy must be minimized with respect to molecular geometry. When the lowest possible energy is reached, the conformation of minimum potential energy in the force field used is obtained.

We shall describe five techniques that have been developed to tackle this problem. A detailed discussion of the first three methods has been given by Williams, Stang and Schleyer [17], therefore we propose to concentrate our discussion on the last method which has become of general use only recently.

1. *Steepest Descent*. This method has been introduced by Wiberg [39] and extensively explored by Allinger *et al.* [40,41,59,60]. Each coordinate in turn is changed by a small amount and the energy change is computed and stored. The coordinate is then returned to its original value and the calculation is repeated for the next one. After the program has cycled over all coordinates, they are changed by an amount proportional to the energy change calculated for each and in the direction which lowers the energy. The same process is repeated until the energy decreases less than a specified amount.

Modifications were introduced in order to save computer time; if symmetry elements are present, only the independent atoms are moved while the symmetry dependent coordinates are adjusted in such a way as to preserve the symmetry. Also during the initial stages of the calculation the hydrogen atoms are not explicitly included but simply follow the atoms to which they are attached. Another time saving procedure is to let the atoms move in tetrahedral directions during the first cycles rather than along the Cartesian axes.

2. *The Parallel Tangents* procedure resembles the steepest descent method. From the starting point two new points (each lower in energy than the previous one) are calculated using a steepest descent technique. The minimum on the curve through these three points yields a fourth point which serves to repeat the procedure until the energy decrease is less than a prespecified amount. An advantage of this method over the steepest descent technique is its scale invariance. [17]

3. *Pattern Search*. This method differs from the steepest descent technique in that a coordinate is left at its more favourable value when the next one is changed. If the initial stepsize is too large to find a lower point on the energy surface (a "pattern"), the stepsize is decreased. This procedure is reported [17] to give lower energies than the steepest descent method and should thus give a better approach of the true minimum.

4. *Non-Simultaneous Local Energy Minimization*. In this scheme, developed by Allinger *et al.* [22], it is assumed that the energy surface near the minimum energy position for each atom can be approximated by

$$V = Ax^2 + By^2 + Cz^2 + Dx + Ey + Fz + G \qquad (14)$$

where xyz are the Cartesian coordinates of the atom local energy. If for two positions of the atom both the potential energy and the partial derivatives of this energy with respect to each Cartesian coordinate are known, the constants $A-F$ of the equation can be calculated. The position of minimum energy is found when the derivative of the equation for the local energy surface is set to zero. In practice, for each atom in turn, the second position is found by a steepest descent technique and the atom is shifted to the minimum energy position which is calculated by the method above. After the program has cycled over all atoms, the process is repeated until the shifts are considered to be sufficiently small.

5. *Valence-Force Minimization.*.This seems thus far the most promising approach because of its reproducibility and versatility. Jacob, Thomson and Bartell [24], Boyd [44] and Lifson and Warshel [19] independently developed computer programs based on iterative minimization of forces ($=$ first derivatives of the energy expressions) by direct solution of the simultaneous linear equations. These programs, besides being efficient in requiring practical amounts of computer time, have the additional advantage that eigenvectors, normal modes of vibrations and thermodynamic functions are readily calculated at the end of the minimization procedure. The description followed here is essentially due to Boyd [44]. The potential energy of the trial model is expanded in a Taylor series around its structural parameters q, where cubic and higher terms are neglected, as:

$$V_{q+\Delta q} = V^0_{q+\Delta q} + \sum_i (\delta V/\delta q_i)\, \Delta q_i + 1/2 \sum_i \sum_j b_{ij}\, \Delta q_i\, \Delta q_j + \cdots \quad (15)$$

with $b_{ij} = \delta^2 V/\delta q_i\, \delta q_j$.

Working in Cartesian coordinates is easier since they are independent parameters whereas the equations of constraint for internal coordinates are difficult to handle. Therefore the Δq have to be transformed to ΔX_i^α ($\alpha = 1,2,3$ represent x,y,z respectively; $i = 1,\ldots,N$). These transformations are made by regarding Δq as a small quantity and expanding in a power series; *e.g.* for Δr_{ij} one obtains:

$$\Delta r_{ij} = \sum_{\alpha=1}^{3} (\delta r_{ij}/\delta X_i^\alpha)_0\, \Delta X_i^\alpha + \sum_{\alpha=1}^{3} (\delta r_{ij}/\delta X_j^\alpha)_0\, \Delta X_j^\alpha +$$
$$+ 1/2 \sum_{\alpha,\beta=1}^{3} \sum_{P,Q} (\delta^2 r_{ij}/\delta X_p^\alpha\, \delta X_q^\beta)_0\, \Delta X_p^\alpha\, \Delta X_q^\beta + \cdots \quad (16)$$
$$\begin{array}{c} (P=I,J) \\ (Q=I,J) \end{array}$$

The derivatives in this equation may be evaluated from the trial Cartesian coordinates, since

$$(r_{ij})_0^2 = (X_j^1 - X_i^1)_0^2 + (X_j^2 - X_i^2)_0^2 + (X_j^3 - X_i^3)_0^2 \quad (17)$$

then

$$(\delta r_{ij}/\delta X_i^a)_0 = -(\delta r_{ij}/\delta X_j^a)_0 = -(X_j^a - X_i^a)_0/r_{ij}^0 \qquad (18)$$

and

$$(\delta^2 r_{ij}/\delta X_i^a \, \delta X_i^\beta)_0 = (\delta^2 r_{ij}/\delta X_j^a \, \delta X_j^\beta)_0 = -(\delta^2 r_{ij}/\delta X_i^a \, \delta X_j^\beta)_0 =$$
$$= \delta^{a\beta}/r_{ij}^0 - (X_j^a - X_i^a)_0 \, (X_j^\beta - X_i^\beta)_0/(r_{ij})_0^3 \qquad (19)$$

where $\delta^{a\beta} = 1$ if $\alpha = \beta$ and $\delta^{a\beta} = 0$ if $\alpha \neq \beta$.

The transformation of structural parameters involving more than two atoms is straightforward, although the complexity of the calculation of the second derivatives forced all workers to use numerical methods rather then analytical formulas. Substitution of the expressions for the Δq_i in Eq. (15) results in an equation for the potential energy that depends only on the ΔX_i^a. From the condition for a stationary value of V,

$$\delta V/\delta X_i^a = 0 \qquad\qquad (\alpha = 1,2,3;\ i = 1,\ldots,N) \qquad (20)$$

a set of $3N$ linear algebraic equations in ΔX_i^a is obtained:

$$-A_i^a = \sum_{j=1}^{N} \sum_{\beta=1}^{3} C_{ij}^{a\beta} \, \Delta X_j^\beta \qquad\qquad (\alpha = 1,2,3;\ i = 1,\ldots,N). \qquad (21)$$

A_i^a is the sum of the coefficients of ΔX_i^a in the linear terms in the expression for the potential energy and $C_{ij}^{a\beta}$ is the sum of the coefficients of ΔX_i^a in the quadratic terms. Matrix C is singular because rotations and translations have not been excluded, therefore, this set of equations cannot be solved directly. The simplest solution is to keep six coordinates constant ($e.g.$ $\Delta x_1, \Delta y_1, \Delta z_1, \Delta x_2, \Delta y_2, \Delta x_3 = 0$) by removing the corresponding rows and columns from the matrix. The resulting set of $3N-6$ equations can now be solved by standard methods.

Since Eq. (15) does not accurately represent the true potential-especially not if the trial model is relatively far away from the true minimum energy geometry — and because of the approximate nature of the transformation of Δq to ΔX, the calculated shifts of the atomic coordinates will not, in general, minimize the potential energy. Therefore the new model is used as input for another cycle of calculations, until the ΔX_i^a are less than a prespecified value. It should be noted that the accuracy of Eq. (16) can be improved by including higher terms; this would probably be outweighted by the increased amount of computer time needed. In the final cycle each A_i^a is very nearly zero and only C-terms remain. The final potential can then be written as:

$$V = V_0 + 1/2 \sum_{i,j=1}^{N} \sum_{a,\beta=1}^{3} C_{ij}^{a\beta} \, \Delta X_i^a \, \Delta X_j^\beta, \qquad (22)$$

Solution of the characteristic equation

$$|C_{ij}^{\alpha\beta} - m_i\, \delta_{ij}\, \omega^2| = 0 \qquad (m_i = \text{mass of } i^{\text{th}} \text{ atom}) \qquad (23)$$

yields the vibrational frequencies $\nu_i = \omega_i/2\pi$ and the eigenvectors (expressed in Cartesian coordinates) of the normal modes of vibration. Together with the moments of inertia as calculated from the final geometry, the frequencies can be used to calculate the rotational and vibrational contributions to the thermodynamic functions.

Symmetry Constraints. Several authors [10,14,26,39] have used constraints to preserve symmetry during calculation. It should be stressed that this is by no means necessary to obtain a good convergence. The greater number of parameters in the absence of symmetry constraints requires a greater amount of computer time and a larger memory capacity. This seeming disadvantage in our view does not outweigh the possible danger of missing double minimum potential wells.

Calculation of Dynamic Properties. The calculation of dynamic changes, *e.g.* calculation of the potential energy curve for the interconversion of two or more conformers, is possible if the interconversion involves a continuous change in one of the internal coordinates and if this coordinate is single valued over the transition path. The "reaction coordinate" can then be determined by constraining this internal coordinate to various values and minimizing the energy with respect to the rest of the geometry. It is very difficult to express such a constraint in terms of Cartesian coordinates, therefore an extra interaction is introduced. This potential has an energy minimum for the desired value of the internal coordinate and must be very stiff to assure that the desired value is closely approached. This extra interaction does not contribute to the total potential energy, it is only included in the calculation of the first and second derivatives which are needed in the minimization procedure. The introduction of such a "constraint" is straightforward for all kinds of internal coordinates except for the torsional potential function, since this has an energy minimum at certain specific values (*e.g.* at 60, 180 and 240 °). This problem has been solved by the introduction of an interaction term [61]

$$V = k(1 + \cos 3(\phi + C)) \qquad (24)$$

or [62]

$$V = k \cos(\phi + C) \qquad (25)$$

where C is a phase angle which shifts the angle of minimum energy to the desired value. This method has been proven to be valuable for the calculation of energy curves for chair-boat interconversions [61], internal rotations [62] and pseudorotation of five-membered rings [62].

Practical Application. As an example of practical application we briefly outline the set-up of an integrated package of computer programs with which the authors have personal experience; its scheme is shown below.

The central routine "UTAH" is a modification of the Fortran "MOLE-CULE BUILDER" program written by Boyd [44]. The input consists of the Cartesian coordinates of the trial model plus a set of interactions, describing the structural parameters and the constants for the chosen potential functions. The trial model is obtained from guessed internal coordinates which are transformed to Cartesians by the routine "FIXAT"[b].

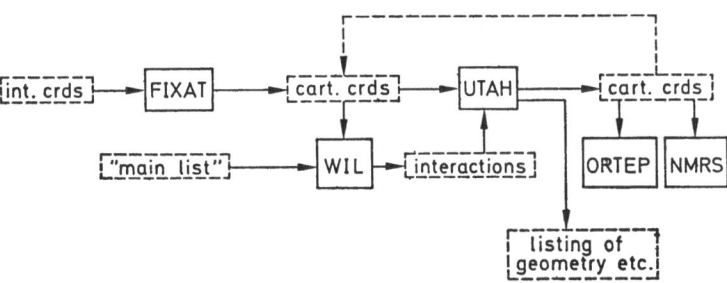

The program "WIL"[b] is a routine which selects the appropriate combinations of atoms and the related parameters. In this scheme each kind of atom is assigned a unique "type number" (*e.g.* hydrogen = 1, primary carbon = 11, etc.). The force constants of every possible interaction are written once (preferably on tape) using these type numbers to denote the participating kinds of atoms. Using this "main list", "WIL" collects the appropriate interactions. An additional feature of this scheme is the automatic assignment of atomic masses to the atoms through the type numbers. The output of "UTAH" consists — besides a list of energies, geometry, vibrational frequencies, eigenvectors and thermodynamic properties — of the Cartesian coordinates of the "refined" model which can be used to restart the procedure. In addition they can be used to visualize the geometry using a modified version of the "ORTEP" [63] program and to calculate the ^1H chemical shifts ("NMRS") by a method similar to that of Tribble *et al.* [64].

V. Comparison of Force Fields

In this section we will try to compare the results obtained when different force fields are applied to some "difficult" molecular systems. It is *not* our intention to qualify these fields as good or bad, but rather to gain some

---

b) Similar modifications of the Boyd program have been described.[65,66]

insight into the factors dominating the balance of forces in the different force fields. Some results were taken directly from the literature (indicated by reference numbers in the corresponding tables), in several other cases we performed the calculations with our own computer programs [67], using published force fields. The force fields employed are: Lifson and Warshell's [21] (LW), Boyd's [26] (B), Altona's [31] (AL) and Allinger's [22] (A) force field. These fields are arranged in order of increasing "hardness" of the hydrogen-hydrogen non-bonded interactions, whereas of course the carbon-carbon non-bonded interactions become correspondently "softer".

(i) *17-β Isopropyl-Androstane* (Fig. 2). This structure was chosen as a model for the steroid skeleton since very accurate experimenal data are now available from a "weighted average" of some 50 X-ray structure deter-

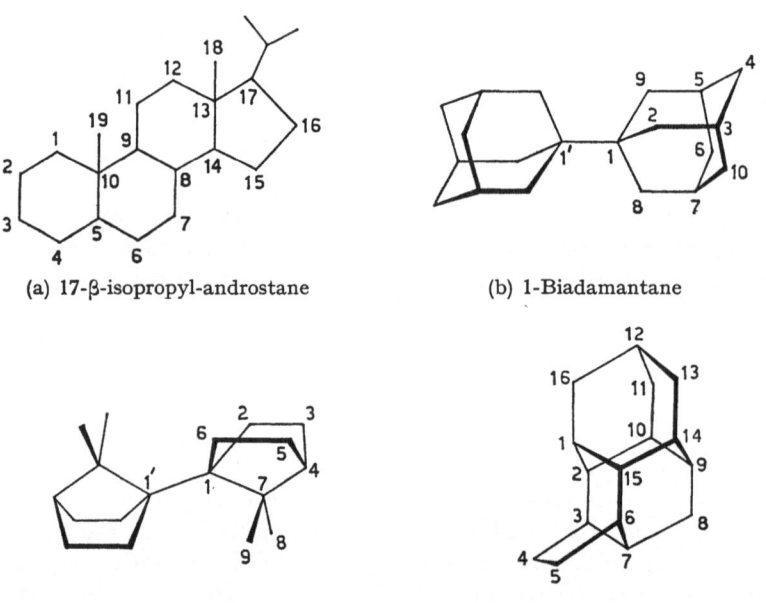

(a) 17-β-isopropyl-androstane

(b) 1-Biadamantane

(c) 1-Biapocamphane

(d) Hexacyclo [10,3,1,0^{2,10},0^{3,7},0^{6,15}, 0^{9,14}]-hexadecane

Fig. 2

minations.[48] In Table 1 the bond lengths resulting from the various force fields are shown, together with the "weighted average" values. It can be seen that all four force fields reproduce the bond distances to within about 0.01 Å (standard deviation), whereas two of them yield standard deviations (0.006 Å and 0.007 Å) that are of a magnitude comparable to a good X-ray analysis. It is also of interest to note that on the whole the bonds are cal-

Table 1. Bond lengths in 17-β-isopropyl-androstane

	Exp.[48]	AL	LW	B	A [70]
1— 2	1.529	1.534	1.523	1.543	1.532
1—10	1.543	1.549	1.544	1.552	1.539
2— 3	1.517	(1.525)[1]	(1.521)[1]	(1.537)[1]	(1.523)[1]
3— 4	1.515	(1.530)[1]	(1.523)[1]	(1.538)[1]	(1.529)[1]
4— 5	1.530	1.538	1.529	1.542	—
5—10	1.553	1.548	1.544	1.557	1.541
5— 6	1.526	1.528	1.528	1.540	1.528
6— 7	1.522	1.531	1.520	1.537	1.527
7— 8	1.532	1.532	1.533	1.541	1.527
8— 9	1.547	1.545	1.549	1.556	1.536
9—10	1.562	1.571	1.541	1.562	1.548
9—11	1.545	1.543	1.533	1.554	1.539
11—12	1.537	1.544	1.533	1.547	1.534
12—13	1.528	1.531	1.515	1.545	1.525
13—14	1.541	1.541	1.563	1.539	1.529
14— 8	1.523	1.532	1.502	1.541	1.528
14—15	1.536	1.523	1.535	1.530	1.533
15—16	1.546	1.546	1.547	1.539	1.540
16—17	1.545	1.554	1.567	1.551	(1.512)[2]
17—13	1.550	1.558	1.540	1.553	(1.514)[2]
13—18	1.538	1.545	1.522	1.553	1.538
10—19	1.546	1.549	1.525	1.552	1.538

Standard deviation

$s = [n^{-1} \Sigma (r\mathrm{exp} - r\mathrm{calc})^2]^{1/2}$

Stand. dev. (Ångstrom):	0.006	0.012	0.011	0.007

[1] These values are excluded from the calculation of the standard deviations since the experimentally determined models contain a substituent at C3.

[2] These values are excluded since in Allinger's calculation C17 was substituted.

culated longer than they actually are. This is especially so for *B* and *AL*, this being understandable in the last case, since this force field was scaled to fit *electron diffraction* data. From Table 2, listing the average torsional angles in the A,B and C rings, it can be seen that the trends are reproduced correctly, and that force fields *LW* and *B* predict the experimental results

Table 2. Average torsional angles in 17-β-isopropyl-androstane

	Exp.[48]	AL	LW	B
Ring A	55.2 ± 0.1	54.2	54.7	55.0
Ring B	55.9 ± 0.1	53.8	56.3	56.3
Ring C	55.5 ± 0.1	53.1	55.9	56.3

pretty accurately. Table 3 shows the actual results from *LW* and *B* compared with the "weighted average" values. It seems surprising that one of the simpler force fields (*B*) is capable of predicting the torsional angles in a rather complicated molecular structure as accurate as to a standard deviation of 1.5°. Together with the calculations to be discussed below, this might indicate that force field *B*, especially its balance between H...H and C...C non-bonded interactions, is a good basis for the construction of future force fields.

Table 3. Torsional angles in 17-β-isopropyl-androstane calculated with two different force fields

	Exp.[48]	B	LW
10— 1— 2— 3	−55.7	−54.5	−54.7
1— 2— 3— 4	53.0	53.1	51.7
2— 3— 4— 5	−53.7	−54.4	−52.8
3— 4— 5—10	56.8	57.5	56.9
4— 5—10— 1	−56.8	−56.1	−56.6
5—10— 1— 2	55.0	54.4	55.7
6— 5—10— 9	58.6	57.9	58.6
10— 5— 6— 7	−58.2	−58.1	−57.6
5— 6— 7— 8	54.2	55.1	54.5
6— 7— 8— 9	−52.5	−54.1	−53.4
7— 8— 9—10	55.0	56.1	55.8
8— 9—10— 5	−56.9	−56.7	−57.6
14— 8— 9—11	−52.6	−51.9	−52.9
8— 9—11—12	52.8	50.4	52.1
9—11—12—13	−54.4	−53.9	−55.7
11—12—13—14	55.7	57.6	56.2
12—13—14— 8	−59.6	−63.2	−59.7
13—14— 8— 9	57.6	60.9	58.5
Standard deviation (degrees)		1.5	0.7

(ii) *1-Biadamantane and 1-Biapocamphane* (Fig. 2). We selected these structures since accurate experimental data were available and because we expected these highly strained systems to reveal some information about the balance between the different interactions. First of all, let us examine some important features of the experimental results (Tables 4 and 5). It is seen that in *1-Biadamantane* (abbreviated as *BAD*) the central carbon-carbon bond is stretched considerably to 1.578 Å, whereas in *1-Biapocamphane* (*BAC*) it is not the central bond but the C_1—C_7 carbon-carbon bond which

Table 4. Symmetry-independent bond lengths in 1-biadamantane

	Exp.[68]	AL	LW	B
1—2	1.546	1.545	1.550	1.558
2—3	1.536	1.536	1.536	1.542
3—4	1.529	1.524	1.535	1.541
4—5	1.529	1.526	1.535	1.541
1—1′	1.578	1.617	1.534	1.581
Stand. dev. (Ångstrom)				
Excluding 1—1′		0.003	0.005	0.011
Including 1—1′		0.011	0.013	0.010

Table 5. Symmetry-independent bond lengths in 1-biapocamphane

	Exp.[68]	AL	LW	B
1—2	1.557	1.555	1.583	1.555
2—3	1.549	1.543	1.546	1.542
3—4	1.527	1.530	1.545	1.540
1—7	1.584	1.578	1.573	1.562
4—7	1.547	1.555	1.581	1.539
7—8	1.535	1.546	1.510	1.550
1—1′	1.544	1.566	1.454	1.570
Stand. dev. (Ångstrom)				
Excluding 1—1′		0.007	0.021	0.012
Including 1—1′		0.009	0.034	0.014

has been stretched to relieve the evidently present strain. From the results of the force fields calculations we see a different picture. Although the differences (excluding the central bond) are still in the order of about 0.01 Å, remarkable differences between both the experimental and calculated values, as well as between the different force fields are revealed. The stretching of the central bond in BAD is overestimated by AL and B, whereas LW even predicts a shortened central bond. In the BAC-case, the central bond length (experimentally about normal) is overestimated by AL and B, but again underestimated by LW, while the lengthened C_1—C_7 bond is predicted more or less by all three fields.

An explanation might be offered by regarding the number of short H...H contacts over the central bond (6 in the BAD case and 4 for BAC) in relation to the "stiffness" of the non-bonded interaction potentials employed in the different fields. The stretching of the central bonds as predicted by the *AL* field is mainly due to these H...H contacts, being described by relatively "hard" potentials, whereas an eventual stretching in the *LW* field would have to be caused by the C—C (1,4) non-bonded interactions (in this field the H...H potentials are relatively "soft"). The overestimation of the central bond lengths by the *AL* field (and, to a certain extent, by *B*) may be ascribed to H—H non-bonded potentials which are too "hard". This of course implies that the corresponding C—C non-bonded interactions are too "soft". The conclusion in the case of the *LW* field might be just the reverse, *i.e.* H—H non-bonded potentials which are too "soft". A better balance seems to have been achieved in *B*, although other factors (such as stretching and 1,3-interactions) may need revision in order to obtain a better fit as far as bond distances are concerned.

(iii) *Hexacyclo* $[10,3,1,0^{2,10},0^{3,7},0^{6,15},0^{9,14}]$-*Hexadecane* (Fig. 2). The MMFF calculations of this interesting ethano-bridged biamantane [69] again show (Table 6) the same trend as noted previously. "Short" carbon-

Table 6. Symmetry-independent bond lengths in hexacyclo $[10,3,1,0^{2,10},0^{3,7},0^{6,15},0^{9,14}]$-hexadecane

	Exp.[69]	AL	LW	B
1— 2	1.534	1.524	1.539	1.549
2—10	1.543	1.534	1.549	1.555
9—10	1.536	1.525	1.544	1.548
1—16	1.524	1.532	1.530	1.542
10—11	1.533	1.531	1.536	1.544
16—12	1.531	1.518	1.533	1.535
11—12	1.525	1.528	1.537	1.543
2— 3	1.545	1.541	1.531	1.560
8— 9	1.516	1.515	1.528	1.530
3— 7	1.531	1.525	1.553	1.534
7— 8	1.517	1.509	1.502	1.534
3— 4	1.541	1.544	1.557	1.544
4— 5	1.552	1.553	1.560	1.552
Stand. dev. (Å)		0.007	0.012	0.012

carbon distances (~ 1.516 Å) are found between C_7—C_8 and C_8—C_9, "long" ones (~ 1.552 Å) between C_4—C_5, C_2—C_{10} and C_2—C_3. The *AL* field follows the trend fairly well, the *LW* and *B* field are second in this respect. It should be noted that the *B* field features a "natural" C—C (r_0) value of 1.53 Å and therefore cannot predict smaller distances than this.

Finally, Table 7 provides a selected bibliography of systems which have been calculated using the MMFF full relaxation method.

Table 7

Alkanes and substituted alkanes	19,21,22,24,26,27,40,41,45,60,71—78)
Medium ring compounds	
unsubstituted	10,19,22,23,28,34,36,39,40,41,45,56,59,61,73,76, 79—82)
alkyl substituted	33,37,40,41,45,56,59,80—84)
functional groups	56,60,82,83)
Fused six-membered ring compounds, bicyclo(4.4.0) type	22,26,40,41,76,80,88—90,115)
Bicyclic ring compounds (n.m.p)	22,31,41,45,73,80,82,85—87),115
Molecules containing functional groups	
amides	20)
carbonyl compounds	56,60,82,91—93)
alcohols	7,56,60,83,86,94)
chlorides	56,60,72,75,95)
cyanides	60)
amines	96)
Molecules containing double bonds	
open chain, 1 C=C	24,25,59,60,97)
cyclic, 1 C=C	8,25,45,59,79,80,91,94,98—100)
cyclic dienes	25,45,101—103)
Steroids	7—9,94)
Aromatic systems	71,104—108)
cyclophanes	32,44)
Hetero atoms in ring skeleton	
silanes	66,109)
ethers, thioethers	110—112)
Coordination complexes	65,113)
Carbonium ions	114)
Interconversion, inversion pathways	25,38,40,59,61,81,91,100)

Acknowledgement. The authors wish to thank the following persons for their valuable aid and cooperation during preparation of this manuscript and especially for carrying out extensive computer calculations: F. J. M. Hoogenboom, C. Blom, S. W. Eisma and H. de Sitter.

C. Altona and D. H. Faber

VI. References

1) Mohr, E.: J. Prakt. Chem. (2) *98*, 315 (1918).
2) Boeseken, J.: Rec. Trav. Chim. Pays-Bas *54*, 101 (1935).
3) Vavon, M. G.: Bull. Soc. Chim. France (4) *49*, 937 (1931).
4) Barton, D. H. R.: Experientia *6*, 316 (1950).
5) Braun, P. B., Hornstra, J., Leenhouts, J. I.: Philips Res. Reports *24*, 441 (1969).
6) Romers, C., Hesper, B., van Heykoop, E., Geise, H. J.: Acta Cryst. *20*, 363 (1966).
7) Altona, C., Hirschmann, H.: Tetrahedron *26*, 2173 (1970).
8) Cohen, N. C.: Tetrahedron *27*, 789 (1971).
9) Allinger, N. L., Wu, F.: Tetrahedron *27*, 5093 (1971).
10) Hendrickson, J. B.: J. Am. Chem. Soc. *83*, 4537 (1961).
11) Lehn, J. M.: Conformational analysis. New York: Academic Press 1971.
12) Hoffman, R.: J. Chem. Phys. *40*, 2745, 2474, 2480 (1964). — Jug, K.: Theoret. Chim. Acta (Berl.) *14*, 91 (1969).
13) Pople, J. A., Segal, G. A.: J. Chem. Phys. *44*, 3289 (1966).
14) Dewar, M. J. S., Haselbach, E.: J. Am. Chem. Soc. *92*, 590 (1970). — Dewar, M. J. S., Kohn, M. C.: J. Am. Chem. Soc. *94*, 2699 (1972).
15) Davidson, R. B., Jorgensen, W. L., Allen, L. C.: J. Am. Chem. Soc. *92*, 749 (1970).
16) Diner, S., Malrieu, J. P., Claverie, P., Jordan, F.: Theor. Chim. Acta (Berl.) *13*, 1 (1969); *15*, 100 (1969).
17) Williams, J. E., Stang, P. J., Schleyer, P. von R.: Ann. Rev. Phys. Chem. *19*, 531 (1968).
18) Morino, Y.: Pure Appl. Chem. *18*, 323 (1969).
19) Lifson, S., Warshel, A.: J. Chem. Phys. *49*, 5116 (1968).
20) Warshel A., Levitt, M., Lifson, S.: J. Mol. Spectry. *33*, 84 (1970).
21) Warshel, A., Lifson, S.: J. Chem. Phys. *53*, 582 (1970).
22) Allinger, N. L., Tribble, M. T., Miller, M. A., Wertz, D. H.: J. Am. Chem. Soc. *93*, 1637 (1971).
23) Bixon, M., Lifson, S.: Tetrahedron *23*, 769 (1967).
24) Jacob, E. J., Thompson, H. B., Bartell, L. S.: J. Chem. Phys. *47*, 3736 (1967).
25) Allinger, N. L., Sprague, J. T.: J. Am. Chem. Soc. *94*, 5734 (1 72).
26) Chang, S., McNally, D., Shary-Tehrany, S., Hickey, M. J., Boyd, R. H.: J. Am. Chem. Soc. *92*, 3109 (1970).
27) Bartell, L. S., Burgi, H. B.: J. Am. Chem. Soc. *94*, 5239 (1972).
28) Dunitz, J. D., Eser, H., Bixon, M., Lifson, S.: Helv. Chim. Acta *50*, 1572 (1967).
29) Bixon, M., Dekker, H., Dunitz, J. D., Eser, H., Lifson, S., Mosselman, C., Sicher, J., Svoboda, M.: Chem. Commun. *1967*, 360.
30) Dunitz, J. D.: Conformations of medium rings. In: Perspectives in structural chemistry II, 1—70 (1968).
31) Altona, C., Sundaralingam, M.: J. Am. Chem. Soc. *92*, 1995 (1970).
32) Shieh, C.-F., McNally, D., Boyd, R. H.: Tetrahedron *25*, 3653 (1969).
33) Hendrickson, J. B.: J. Am. Chem. Soc. *84*, 3355 (1962).
34) Hendrickson, J. B.: J. Am. Chem. Soc. *86*, 4854 (1964).
35) Hendrickson, J. B.: J. Org. Chem. *29*, 991 (1964).
36) Hendrickson, J. B.: J. Am. Chem. Soc. *89*, 7036 (1967).
37) Hendrickson, J. B.: J. Am. Chem. Soc. *89*, 7043 (1967).
38) Hendrickson, J. B.: J. Am. Chem. Soc. *89*, 7047 (1967).
39) Wiberg, K. B.: J. Am. Chem. Soc. *87*, 1070 (1965).
40) Allinger, N. L., Miller, M. A., VanCatledge, F. A., Hirsch, J. A.: J. Am. Chem. Soc. *89*, 4345 (1967).
41) Allinger, N. L., Hirsch, J. A., Miller, M. A., Tyminski, I. J., VanCatledge, F. A.: J. Am. Chem. Soc. *90*, 1199 (1968).

36

42) Westheimer, F. H.: J. Chem. Phys. *14*, 733 (1946).
43) Westheimer, F. H.: Steric effects in organic chemistry, Chap. 12. New York: Wiley 1956.
44) Boyd, R. H.: J. Chem. Phys. *49*, 2574 (1968).
45) Schleyer, P. v. R., Williams, J. E., Blanchard, K. R.: J. Am. Chem. Soc. *92*, 2377 (1970).
46) Burgi, H. B., Bartell, L. S.: J. Am. Chem. Soc. *94*, 5236 (1972).
47) Simanouti, T.: J. Chem. Phys. *17*, 245, 848 (1949).
48) Romers, C., Altona, C., Jacobs, H. J. C., De Graaf, R. A. G.: to be published.
49) Bartell, L. S.: J. Chem. Phys. *32*, 827 (1960).
50) Snyder, R. G., Schachtschneider, J. H.: Spectrochim. Acta *21*, 169 (1965).
51) Bartell, L. S.: Abstract 5.31, 6th Int. Congress Internatl. Union of Crystallography (1963).
52) Adams, W. J., Geise, H. J., Bartell, L. S.: J. Am. Chem. Soc. *92*, 5013 (1970).
53) Dobler, M., Dunitz, J. D., Mugnoli, A.: Helv. Chim. Acta *49*, 2492 (1966).
54) Williams, D. E.: J. Chem. Phys. *45*, 3770 (1966).
55) Williams, D. E.: J. Chem. Phys. *47*, 4680 (1967).
56) Altona, C., Sundaralingam, M.: Tetrahedron *26*, 925 (1970).
57) Blom, C. E., Altona, C.: to be published.
58) Hill, T. L.: J. Chem. Phys. *16*, 399 (1948).
59) Allinger, N. L., Hirsch, J. A., Miller, M. A., Tyminski, I. J.: J. Am. Chem. Soc. *90*, 5773 (1968).
60) Allinger, N. L., Hirsch, J. A., Miller, M. A., Tyminski, I. J.: J. Am. Chem. Soc. *91*, 337 (1969).
61) Wiberg, K. B., Boyd, R. H.: J. Am. Chem. Soc. *94*, 8426 (1972).
62) Faber, D. H., Altona, C.: to be published.
63) Johnson, C. K.: Chem. Div. Ann. Progr. Rept. 4164, p. 116, Oak Ridge National Laboratory, Oak Ridge, Tennessee (1967).
64) Tribble, M. T., Miller, M. A., Allinger, N. L.: J. Am. Chem. Soc. *93*, 3894 (1971).
65) Buckingham, D. A., Maxwell, I. E., Sargeson, A. M., Snow, M. R.: J. Am. Chem. Soc. *92*, 3617 (1970).
66) Ouelette, R. J., Baron, D., Stolfo, J., Rosenblum, A., Weber, P.: Tetrahedron *28*, 2163 (1972).
67) Faber, D. H.: Forthcoming Thesis, Leiden.
68) Alden, R. A., Kraut, J., Traylor, T. G.: J. Am. Chem. Soc. *90*, 74 (1968).
69) Rao, S. T., Sundaralingam, M.: Acta Cryst. *B 28*, 694 (1972).
70) Allinger, N. L., Wu, F.: Tetrahedron *27*, 5093 (1971).
71) Kitaigorodskii, A. I., Dashevskii, V. G.: Theoret. Exp. Chem. *3*, 18, 22 (1967).
72) Abraham, R. J., Parry, K.: J. Chem. Soc. B *1969*, 539.
73) Kitaigorodskii, A. I.: Tetrahedron *14*, 230 (1961).
74) Heublein, G., Ruehmstedt, R., Kadura, P., Dawczynski, H.: Tetrahedron *26*, 81 (1970).
75) Heublein, G., Ruehmstedt, R., Dawczynski, H., Kadura, P.: Tetrahedron *26*, 91 (1970).
76) Geneste, P., Lamaty, G.: Bull. Soc. Chim. France *1967*, 4456.
77) Abe, A., Jernigan, R. L., Flory, P. J.: J. Am. Chem. Soc. *88*, 631 (1966).
78) Boyd, R. H., Breitling, S. M.: Macromolecules *5*, 1 (1972).
79) Bucourt, R., Bull. Soc. Chim. France *1964*, 2080.
80) Bucourt, R., Hainaut, D.: Bull. Soc. Chim. France *1965*, 1366.
81) Schmid, H. G., Jaeschke, A., Friebolin, H., Kabuss, S., Mecke, R.: Org. Magn. Res. *1*, 163 (1969).
82) Fournier, J., Waegell, B.: Tetrahedron *26*, 3195 (1970).

83) Faber, D. H., Altona, C.: Chem. Commun. *1971*, *1210*.
84) Allinger, N. L., Pamphilis, N. A.: J. Org. Chem. *36*, 3437 (1971).
85) Ermer, O., Dunitz, J. D.: Helv. Chim. Acta *52*, 1861 (1969).
86) Coulombeau, C., Rassat, A.: Tetrahedron *28*, 2299 (1972).
87) Fournier, J., Waegell, B.: Tetrahedron *28*, 3407 (1972).
88) Bucourt, R.: Bull. Soc. Chim. France *1963*, 1262.
89) Bucourt, R., Hainaut, D.: Bull. Soc. Chim. France *1966*, 501.
90) Allinger, N. L., Gordon, B. J., Tyminsky, I. J., Wuesthoff, M. T.: J. Org. Chem. *36*, 739 (1971).
91) Buccourt, R., Hainaut, D.: Bull. Soc. Chim. France *1967*, 4562.
92) Allinger, N. L., Tribble, M. T.: Tetrahedron *28*, 1191 (1972).
93) Allinger, N. L., Tribble, M. T., Miller, M. A.: Tetrahedron *28*, 1173 (1972).
94) Portheine, J. C.: Thesis, Leiden (1971).
95) Goussot-Leray, A., Bodot, H.: Tetrahedron *27*, 2133 (1971).
96) Allinger, N. L., Hirsch, J. A., Miller, M. A.: Tetrahedron Letters *38*, 3729 (1967).
97) Eisma, S. W., Altona, C., Geise, H. J., Mijlhof, W., Renes, G. H.: J. Mol. Struct., in the press.
98) Bucourt, R., Hainaut, D.: Compt. Rend. Acad. Sci. Paris *1964*, 3305.
99) Favini, G., Buemi, G., Raimondi, M.: J. Mol. Struct. 2, 137 (1968).
100) Dashevskii, V. G., Naumov, V. A., Zapirov, N. M.: J. Struct. Chem. *11*, 687 (1970).
101) Favini, G., Zuccarello, F., Buemi, G.: J. Mol. Struct. *3*, 385 (1969).
102) Buemi, G., Favini, G., Zuccarello, F.: J. Mol. Struct. *5*, 101 (1970).
103) Allinger, N. L., Tribble, M. T., Sprague, J. T.: J. Org. Chem. *37*, 2423 (1972).
104) Allinger, N. L., Sprague, J. T., Finder, C. J.: Abstr. of papers 161st Natl. Meeting A.C.S., 1971, Abstr. Orgn. 05.
105) Archer, R. A., Boyd, D. B., Demarco, P. V., Tyminski, I. J., Allinger, N. L.: J. Am. Chem. Soc. *92*, 5200 (1970).
106) Mannschreck, A., Ernst, L.: Chem. Ber. *104*, 228 (1971).
107) Allinger, N. L., Maul, J. J., Hickey, M. J.: J. Org. Chem. *36*, 2747 (1971).
108) Allinger, N. L., Tribble, M. T.: Tetrahedron Letters *1971*, 3259.
109) Tribble, M. T., Allinger, N. L.: Tetrahedron *28*, 2147 (1972).
110) Seip, H. M.: Acta Chem. Scand. *23*, 2741 (1969).
111) Náhlovska, Z., Náhlovsky, B., Seip, H. M.: Acta Chem. Scand. *23*, 3554 (1969).
112) Náhlovska, Z., Náhlovsky, B., Seip, H. M.: Acta Chem. Scand. *24*, 1903 (1970).
113) Snow, M. R.: J. Am. Chem. Soc. *92*, 3610 (1970).
114) Gleicher, G. J., Schleyer, P. v. R.: J. Am. Chem. Soc. *89*, 582 (1967). — Schleyer, P. v. R., Isele, P. R., Bingham, R. C.: J. Org. Chem. *33*, 1239 (1968). — Bingham, R. C., Sliwinski, W. F., Schleyer, P. v. R.: J. Org. Chem. *92*, 3471 (1970). — Bingham, R. C., Schleyer, P. v. R.: J. Am. Chem. Soc. *93*, 3189 (1971); Tetrahedron Letters *1971*, 27. — Karim, A., McKervey, M. A., Engler, E. M., Schleyer, P. v. R.: Tetrahedron Letters *1971*, 3987. — Fry, J. L., Engler, E. M., Schleyer, P. v. R.: J. Am. Chem. Soc. *94*, 4628 (1972).
115) Engler, E. M., Chang, L., Schleyer, P. v. R.: Tetrahedron Letters *1972*, 2525. — Gorrie, T. M., Engler, E. M., Bingham, R. C., Schleyer, P. v. R.: Tetrahedron Letters *1972*, 3039. — Engler, E. M., Blanchard, K. R., Schleyer, P. v. R.: Chem. Commun. *1972*, 1210.

Received June 13, 1973

The Removal of Orbital Symmetry Restrictions to Organic Reactions

Frank D. Mango

Shell Development Company, Houston, Texas

Contents

I. Introduction

Molecular orbital symmetry conservation focuses attention on the number of bonds that are preserved throughout a chemical transformation [1]. In a symmetry-allowed reaction, the number of bonds remains fixed across the reaction coordinate. Although this is true for all allowed reactions, significant variations can exist in the *degree* of bonding that is preserved. An allowed reaction may suffer sufficient net loss in total bond order approaching the transition state to encounter a formidable energy barrier. Reactions, therefore, classified symmetry-allowed need not proceed at observable rates.

Although a symmetry-allowed assignment places no obligation to react on a molecular system, a symmetry-forbidden assignment, in most cases, precludes reaction. It is in this category of organic reactions that the symmetry rules exhibit their constrictive properties. A molecular system proceeding along a forbidden path of transformation experiences, to a first approximation, the net loss of a full bond approaching the transition state. This somewhat simplified picture of the forbidden process is adopted here to draw attention to the importance of bond preservation, for it constitutes the focal point for understanding the catalysis of symmetry-forbidden reactions. Molecular systems will reasonably tend to react along paths in which bond preservation is maximized. When there is severe loss of bonding, high energies of activation are to be anticipated and reactions along these paths will be infrequent. Chemical agents can intervene which restore bonding along the reaction coordinate, thereby catalyzing the reaction. The restoration of bonding to a symmetry-forbidden reaction is a special case and is the subject of this chapter.

Symmetry-forbidden transformations are rarely observed in organic chemistry. However, a variety of transition metal complexes dramatically catalyze them [2]. The nature of this catalysis is undoubtedly complex and quite diverse and has only recently become the subject of serious interest. Two kinds of mechanisms have been considered. At the extreme, the transition metal removes all symmetry-restrictions to the reaction of its ligands [3]. In this process, termed "forbidden-to-allowed", the metal, using the unique symmetry properties of its d orbitals, restores bonding to the transforming ligand system by injecting its valence electrons into ligand orbitals and withdrawing ligand electrons from incipient antibonding orbitals. This is no incidental act by the metal; it springs from the orbital components of the coordinate bond and is directly associated with the extent and preservation of coordinate bonding. The forbidden-to-allowed catalytic process is marked by the concerted character of the ligand transformation. That is, the ligand transformation viewed alone exactly mirrors the metal-free, symmetry-forbidden organic reaction.

The other category of catalytic mechanisms contains the so-called nonconcerted processes. The term "nonconcerted" may have been unfortunately attached to this chemistry, since it implies freedom from symmetry control. Each step in an overall reaction, however, should be treated as a concerted transformation, subject to symmetry restrictions. Nevertheless, this category of reactions addresses distinctly stepwise processes characterized by intermediates between reacting and product systems. The ligand transformation, viewed alone, bears no resemblance to the metal-free, symmetry-forbidden reaction. Stepwise reactions constitute a broad area in catalysis. Consider, for example, metal-catalyzed isomerization of a carbon-carbon double bond. Double bond migration is readily achieved through the generation of distinct intermediates such as π-allyl or metal alkyl species [4]. Each of these paths differs from the concerted in that the coordinate bond attaching the olefin substrate to the metal undergoes a distinct change in character upon generating the intermediate. Moreover, the number of atoms in the rearranging ligand changes in the critical step. In the π-allyl path, a hydrogen atom is extracted from the ligand by the metal; in the metal alkyl path, a hydrogen atom is donated to the olefin ligand by the metal.

Stepwise paths as described here are clearly distinct from the concerted, forbidden-to-allowed process, and this distinction remains firm so long as intermediates are not considered transition states. A transition state resembling a π-allyl intermediate has been suggested [5] for the iron carbonyl-catalyzed isomerization of the *endo*alcohol *1* to ketone *2* [6].

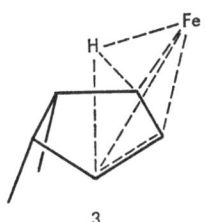

In the proposed mechanism, transition state *3* is introduced, thus avoiding intervention of a distinct π-allyl intermediate.

3

There is, however, no evidence indicating that a π-allyl intermediate does not intervene. The stepwise process, having broader precedence, would seem to be the preferred in the absence of compelling evidence to the contrary.

II. Ligand-Metal Interaction

In catalysis of organic reactions by transition metals, the organic molecular system is fused to the inorganic. Carbon s and p orbitals mix with metal s, p, and d orbitals, and metal valence electrons penetrate the organic network of bonds. A new molecular species is created prossessing distinctly different molecular degrees of freedom. An organic molecule, upon coordination to a metal, may have reaction paths open to it which were not available prior to coordination. Although the reaction behavior of a metal-coordinated organic system may differ sharply from that of the pure organic system, it is similarly controlled by orbital symmetry principles. However, it is a different molecular species, and thus will experience a different set of transformation constraints.

We wish to address this subject somewhat generally. There are, conceivably, a number of ways it can be approached. The metal complex can, for example, be treated as an inorganic system with attention focused on the distribution of metal valence electrons, their spin multiplicities and the relationship between these factors and the dynamics of the metal's ligands. Or the focal point can be the reaction of the ligands and the specific demands imposed on the metal by this process. We shall tend to follow this latter line. Most of the reactions to be discussed are symmetry-forbidden and thus require clear operations by the metal. To simplify the picture further, we shall, at times, treat the metal as a pseudo-organic moiety attached to an organic system. This will allow a closer analogy to the broad treatment of pure organic reactions recently published [1].

First, we shall examine the bonding properties of a metal center having d orbitals within its valence band and nonbonding valence electrons. Consider the interaction between this metal and a single bond within some organic molecule.

$$\overset{\displaystyle \ \ ..}{\underset{4}{M}} \ \overline{\quad\quad\quad} \quad \longrightarrow \quad \underset{5}{\bigvee_{M}} \tag{1}$$

Eq. 1 is a simple valence bond representation of this process. The bond (represented by the horizontal line in 4) is spatially relocalized by the metal. Structures 4 and 5 reflect the shift in bonding patterns. A particular bonding character (*i.e.*, π or σ)need not be assumed from these structures. We shall show later that clear distinctions between simple coordination (π bonding) and full σ bonding (as implied in *5*) are often difficult to draw. For our purposes, no distinction need be drawn; *5* can either represent simple coordination or complete oxidative addition.

Consider *4 → 5* a [2+2] cycloaddition process in which the metal is a pseudo-organic participant. The electron pair in the organic moiety rests in a symmetric orbital (relative to a plane passing through the metal and bisecting the bond axis); the metal must therefore donate its valence electrons through an antisymmetric orbital if this is to be an allowed $[2_s + 2_a]$ transformation. This process becomes clearer if viewed as a combination of metal and ligand orbitals. The molecular orbital description of the bonding network in *5* contains the following molecular orbitals with the indicated symmetries (S = symmetric and A = antisymmetric relative to the mirror plane).

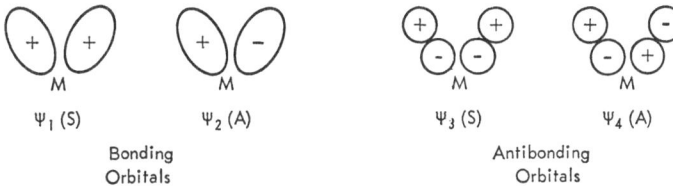

Ψ_1 (S) Ψ_2 (A) Ψ_3 (S) Ψ_4 (A)

Bonding   Antibonding
Orbitals   Orbitals

This orbital representation illustrates only the symmetry pattern for this bonding configuration. The orbitals are therefore displayed as simple elliptical figures of the appropriate symmetries superimposed over the bonding regions. Other orbital approximations can be constructed from linear combinations of atomic *s*, *p*, and *d* orbitals. Although they would differ in shape from these, their symmetry properties would be identical. Throughout this chapter we shall attempt to adopt the simplest orbital representations consistent with the points at hand.

In bonding network *5*, Ψ_1 and Ψ_2 are occupied with electrons and Ψ_3 and Ψ_4 are empty. To form *5* from *4* (Eq. 1), the participants in *4* (*i.e.*, the metal and the incipient ligand) must provide orbitals with these symmetries, and they must be appropriately occupied. The proper combinations are illustrated in Fig. 1.

The metal, therefore, provides two atomic orbitals, one symmetric and empty and the other antisymmetric and populated, in order to interact in a bonding way with a bond having the symmetry properties indicated in *4*. The energy and symmetry properties of the *d* orbitals are ideal for this. In coordinate system *6*, the metal's atomic orbitals fall into the two categories with respect to the ZY plane as follows: symmetric (s, p_y, p_z, d_{z^2}, $d_{x^2-y^2}$, d_{zy}); antisymmetric (p_x, d_{xy}, d_{zx}). Thus, *d* orbitals of both symmetries are available. Moreover, the interaction of the incoming ligand with the metal would tend to order the d orbitals as indicated in Fig. 1.

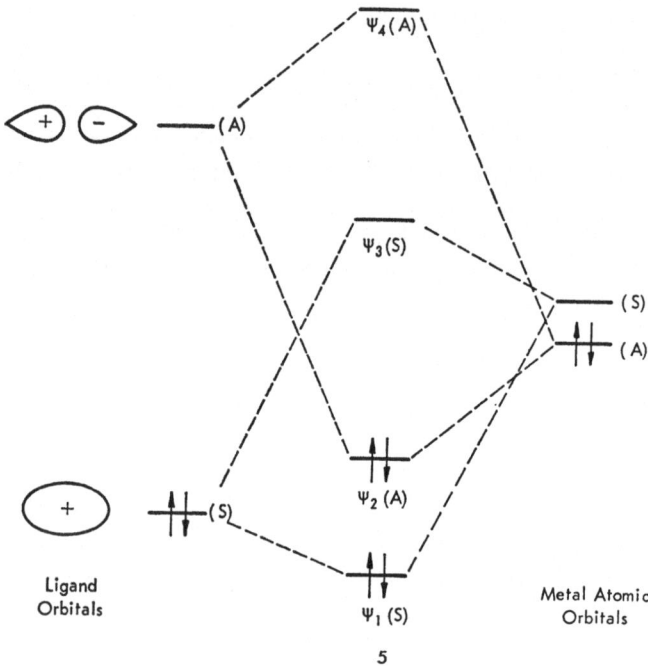

5

Fig. 1. The orbital combinations for a symmetry-allowed [2 + 2] metal-ligand combination

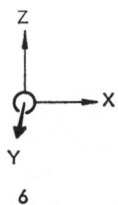

6

This situation should be contrasted to pure organic systems where orbitals with correct symmetries are available for suprafacial [2 + 2] cyclo-additions, but their electronic occupations are inappropriately fixed by large energy gaps (on the order of several electron volts) separating the bonding from antibonding. The transition metal, in contrast, possesses d orbitals of both symmetries capable of assuming a broad variety of electronic occupations depending on the number of valence electrons and ligand field factors. The metal can thus be looked upon as a very versatile pseudo-organic participant capable of providing and accepting electron pairs in almost any required symmetry configuration. The capacity of transition metal complexes to coordinate the variety of organic compounds (olefins, acetylenes, dienes, trienes, etc.) rests upon the symmetry and energy properties of their d orbitals. The special catalytic properties of the transition elements stem, to a large degree, from their ability to adopt a spectrum of bonding configurations by using different combinations of d orbitals and valence electrons.

The symmetry factors just described for the $[2+2]$ cycloaddition process are general and apply equally to simple bond coordination and complete bond fusion (oxidative addition). The interacting bond can be either σ or π; only its symmetry is critical. The σ bond in a cyclopropane ring, for example, can assume a $[2+2]$ interaction with a metal center to any point between the two extremes of simple coordination (7) and complete oxidative addition (8). In either situation, the symmetries of the bonding networks are

7   8

the same, assuming normal donor and back-bonding interactions in 7. The metal's capacity to form a simple coordinate bond to a monodentate ligand is thus closely related to its ability to completely cleave a bond, totally separating the previously bonded nuclei. The orbital overlap and energy properties of the system in question will dictate where along the continuum between the extreme points a system will actually rest.

The metal's role in cycloaddition processes can be extended beyond $[2+2]$. The symmetry of the atomic orbital used by the metal to donate or withdraw the electron pair will be dictated by the Woodward-Hoffmann symmetry rules [1]. As in the $[2+2]$ case, the metal is looked upon as a pseudo-organic participant. When an orbital reacting in a suprafacial way is called for by the symmetry rule, a metal symmetric orbital is used. Similarly, when an antarafacial participant is called for, a metal antisymmetric orbital is used. In the $[2+2+2]$ cycloaddition process, for example, where the ligand participants must react along the suprafacial mode, the

$$ \qquad\qquad\qquad\qquad\qquad (2) $$

metal contributes an electron pair through a symmetric orbital to yield the symmetry-allowed cycloaddition.

Higher orders of reaction are treated similarly. Once again, the restrictions found in organic chemistry are not to be as generally encountered when the metal participates. [2+2] cycloaddition processes, for example, are rare in olefin chemistry since one of the participants must use its reacting π bond in an antarafacial manner to effect reaction along the allowed $[_\pi 2_s + _\pi 2_a]$ path. This places strong geometrical restrictions on olefin participants which very often preclude reaction. The transition metal, because it contains valence electrons and a variety of symmetric and antisymmetric orbitals within a narrow energy band, participates in [2+2] processes freely.

As we proceed with our treatment of actual catalytic processes, it may be best to adopt the notations introduced by Woodward and Hoffmann in their generalized selection rule for pericyclic reactions [1]. We shall identify the metal participant by the subscript "d", referring to the d orbital. Exclusion of participation by metal s and p orbitals is not implied. The subscripts "s" and "a" will refer to symmetric and antisymmetric, respectively, when attached to the metal participant. This is to avoid the ambiguity associated with the concepts "suprafacial" and "antarafacial" when applied to d orbitals. Consider the transformation in Eq. 1, for example. In coordinate system 6, the symmetric atomic orbitals available to the metal participant are the s, p_y, p_z, d_{z^2}, $d_{x^2-y^2}$ and d_{zy}; the remaining three atomic orbitals are antisymmetric. In the symmetry-allowed transformation, the metal participant is thus represented $_d 2_a$, indicating a metal center (d) with two electrons participating in the reaction and using one of its antisymmetric (a) atomic orbitals (*i.e.*, its p_x, d_{xy} or d_{zx}). The [2+2] cycloaddition reaction itself, then, will be described $[_d 2_a + _\pi 2_s]$ for a ligand with a π bond participating in the reaction and $[_d 2_a + _\sigma 2_s]$ for a ligand with a σ bond interacting. The transformation in Eq. 2 is similarly designated $[_d 2_s + _\pi 2_s + _\pi 2_s]$. The reverse reaction in Eq. 2 will proceed along the $[_\sigma 2_s + _\sigma 2_s + _\sigma 2_s]$ path. The metal participant is not specifically indicated, but it is clear from the reaction that the metal must withdraw an electron pair from the transforming ligand system into one of its symmetric (s) orbitals.

It is doubtful that orbital symmetry conservation will maintain pronounced constrictive control over cycloaddition processes when transition metals participate. Certainly, the absolute order reflected in the sharp division between allowed and forbidden processes found in organic chemistry would not be anticipated in this special area. The reasons are primarily those just discussed. But the importance of orbital symmetry control does not rest on its predictive powers alone. It provides insight into the nature of a molecular transformation, focusing attention on the character of transforming bonds as reflected in the symmetries of their composite molecular orbitals. The facility with which various metal systems participate in cycloaddition processes should be closely associated with the positioning of particular *d* orbitals within the valence band and their electron occupation. Simple symmetry principles, then, might offer the inorganic chemist a clearer view of fundamental processes in catalysis since they highlight the critical orbitals.

III. Classes of Ligand Transformations

Our interest will be centered primarily on the metal's role in catalyzing symmetry-forbidden reactions. This, as we shall point out, is a special case and one that might best be introduced in contrast to other kinds of metal-assisted ligand transformations. Pursuing this approach, we shall consider generally the transformation of a ligand A to some ligand B.

First we shall treat the situation in which the ligand transformation A →B is, by itself, symmetry-allowed (Class 1).

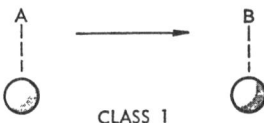

The ligand reaction A → B, then, exactly mirrors the allowed transformation in a pure organic reaction. Examples include the symmetry-allowed Cope rearrangement $[_\sigma2_s + _\pi2_s + _\pi2_s]$ (Eq. 3), the Diels-Alder cycloaddition $[_\pi4_s + _\pi2_s]$ (Eq. 4) and conrotatory ring-opening of cyclobutene $[_\pi2_s + _\sigma2_a]$

$$(3)$$

$$(4)$$

$$(5)$$

(Eq. 5). The essential feature of a symmetry-allowed reaction is the symmetry-match of occupied molecular orbitals in A with those in B and a similar match of unoccupied A and B orbitals. Thus a band of bonds (occupied molecular orbitals) exists in A corresponding in symmetry to a similar band in B. The transition metal, upon coordinating A or B, will mix its set of atomic orbitals with the occupied and unoccupied sets of the ligand

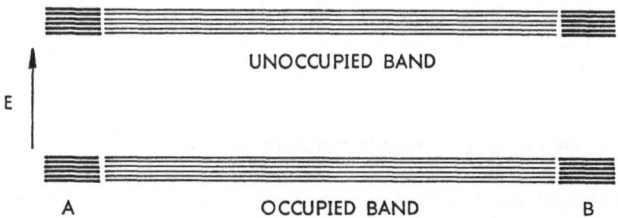

Fig. 2. A general description of a symmetry-allowed molecular transformation

system. The composite will represent the coordinate bond. The ligand's orbitals, considered separately, will have been shifted in energy and altered in net electronic occupation due to metal interaction, but their basic pattern as described in Fig. 2 will not have been significantly changed, in most cases. The capacity of a ligand A to transform to a ligand B rests in the symmetry-match illustrated in Fig. 2, and this should be retained upon coordination to the metal. Stated another way, the ligand itself (A) carries the bonding network (occupied molecular orbitals) which describes the bonding network of the product ligand (B). The ligand transformation, therefore, viewed alone, should remain symmetry-allowed; the metal, in most cases, should not inject into the ligand system sufficient perturbation to effect a net reordering of ligand occupied and unoccupied molecular orbitals so that a genuine crossing of ligand occupied and unoccupied orbitals occurs. Exceptions might occur when the energy between bands is small.

Significantly, the metal is not a participant in the A-to-B transformation in the sense discussed above. The bonds describing B generate from A, not from the metal system. The metal's role in this transformation, however, need not be entirely passive. At one extreme, it can serve as a template, placing ligands within bonding proximity, as in the trimerization of olefin ligands $[_\pi 2_s + _\pi 2_s + _\pi 2_s]$. But the metal loses the coordinate bonding associated with the three π bonds in this A → B process. With three acetylenes, coordinate bonding would be preserved, and this transformation would be a predictably more favorable catalytic process. The preservation of coordinate bonding in the acetylene case stems from the fact that the basic distribution (ordering) of metal valence electrons does not generally change in the Class 1 reaction. If the metal's valence electrons are spatially distributed to coordinate-bond to system A (Eq. 6), they will be properly distributed to bond to system B, since the tridentate centers within the ligand system have not changed.

$$\underset{A}{\text{[structure]}} \quad \longrightarrow \quad \underset{B}{\text{[structure]}} \qquad (6)$$

This will not prevail when A → B constitutes a symmetry-forbidden trans-formation (*i.e.*, Class 2), as we shall demonstrate later. Class 1 and 2 reactions are sharply contrasted in this respect.

Class 1 reactions can be more complex. We have noted that coordinate bonding can be lost due to the disappearance of π bonds in the $[_{\pi}2_s + _{\pi}2_s + _{\pi}2_s]$ process and the general feature that metal valence electrons need not be redistributed with A → B. These are, however, the simplest cases. The ligand centers of coordination can shift spatially with A → B, as in the Cope rearrangement (Eq. 3). This should not impede reaction, however, since the metal center is capable of retaining coordination by shifting with the moving bond system. Other transformations are more complex in their transformed array of coordinate bonding centers. This is true for ligand systems which undergo changes in coordination numbers like the Diels-Alder reaction (tridentate → monodentate, Eq. 4) and conrotatory cyclobutene ring opening (monodentate → bidentate, Eq. 5).

The ordering of metal valence electrons describing a coordinate bond to A need not be that required for B. In these circumstances, the number of metal valence electrons can play an important role in the extent of coordinate bonding that is retained in the A → B transformation. This chemistry can be complicated by orbital crossings involving primarily d levels, which place certain metal complex reactions formally in symmetry-forbidden categories. In these cases, the ligand transformation (A → B) may be considered allowed; ligand system B, however, presents an array of coordinate bonding centers which interacts unfavorably with the metal center. This aspect of Class 1 reactions, however, is a separate subject and will not be treated here. For our purposes, we shall consider these ligand transformations "allowed", keeping in mind that some metal systems may actually impede reaction through negative bonding to system B.

Our primary interest in Class 1 reactions is to contrast symmetry-forbidden ligand transformations. In this regard, the important feature to Class 1 processes is that the metal does not formally participate in the ligand pericyclic process. Moreover, the metal need not undergo a redistribution of its valence electrons and the ligand transformation A → B, viewed alone, remains symmetry-allowed. The catalytic role of the metal in this process, and the negative or positive effects of metal-ligand interaction on the activation energy of A → B, are more appropriately treated separately [29].

The second class of reactions (Class 2) contains those processes in which the ligand transformation A → B is, by itself, symmetry-forbidden. This is the forbidden-to-allowed process; it requires special operations on the part of the metal which place it in a class by itself. For the cycloaddition reactions that we will be concerned with, the metal exchanges a pair of electrons with the transforming ligand and, in the process, suffers a spatial redistribution of its valence electrons. In this instance, Class 2 reactions are represented by

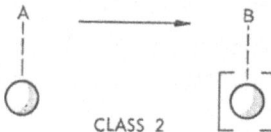

The brackets around the metal are used to designate a different ordering of metal valence electrons at that point of reaction. It will be shown later that ligand-to-metal coordinate bonding is preserved in this process. That is, the reordering experienced by the metal is such that it corresponds to that required for a full coordinate bond to B. The redistribution of metal valence electrons, however, can introduce special energy barriers to reaction. Transformation A → B can generate electron density within the metal complex which creates antibonding character with either the nonreacting portions of B (the residual π bonds in a *bis*acetylene system, for example) or the nonreacting ligands attached to the metal. Both situations will be treated.

The third class of ligand transformations that we wish to distinguish contains those transformations in which the metal serves as a participant in the formation of a new bonding configuration. The bonds between the metal

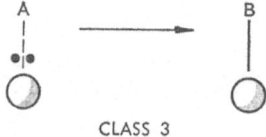

and ligand, in these cases, cannot be clearly distinguished in character from those within the ligand system. The metal, in other words, becomes locked into the bonding configuration of the ligand system. We shall be concerned with those cases where ligand system A is bonded to the metal in a way different from that of B. A can be a free species entering the coordination sphere or a coordinated ligand undergoing an oxidative process with the metal. The transformation in Eq. 2 is an example. As noted in the previous section, a considerable grey area exists between simple coordination as in 7 and actual oxidative insertion [8]. These distinctions become important when we discuss the removal of symmetry restrictions to a reaction by concerted and stepwise processes. Clear distinctions between them may not always exist. Class 3 reactions are generally distinguished from Class 1 in that the transformation A → B (Class 3) requires participation of the metal and has no counterpart in the chemistry of the free and isolated system A. They differ from Class 2 reactions, of course, in that A → B (Class 2) is, by itself, symmetry-forbidden. Class 3 reactions follow the simple symmetry rules described in the previous section and play an important role in the catalysis of symmetry-forbidden reactions.

IV. The Catalysis of Symmetry-Forbidden Reactions

A symmetry-forbidden reaction can be switched to an allowed reaction in a number of ways. One of the more interesting mechanisms is that in which the actual forbidden transformation takes place on the coordination sphere of a transition metal. The ligand transformation here is concerted, the symmetry restrictions having been removed by the metal. The metal's role in this process has been described briefly in an earlier communication by this author with J. H. Schachtschneider [3], and in more detail in a broader treatise [2]. The description will not be repeated here; instead, the subject will be approached from a different point of view, one that focuses attention on the coordinate bond and its relationship to the forbidden-to-allowed process.

A. The Role of the Coordinate Bond

When an olefin coordinates to a transition metal, the olefin π bond donates electrons to an empty metal orbital (donor bond) and the olefin π* orbital accepts metal valence electrons from a filled metal atomic orbital (backbond). Two molecular orbitals can describe the conventional representation of the metal-olefin bond as originally proposed by Dewar and modified by Chatt [7].

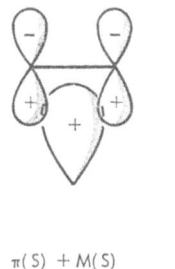
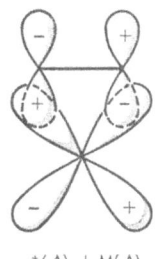

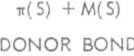

π(S) + M(S)   π*(A) + M(A)

DONOR BOND   BACK-BOND

In this representation, we need not specify which orbital or combination of orbitals the metal provides, only the symmetry match is important. Both donor and back-bonding interactions tend to reduce the bond order of the coordinated ligand. In the donor bond, π bonding electron density shifts to the metal. In the back-bond, the ligand's excited state is partially populated by metal valence electrons. This tends to reduce carbon-carbon bond order, altering the structure of the ligand in the direction of its excited state. The degree of back-bonding is reflected in carbon-carbon and metal-olefin stretching frequencies in the infrared region, and also in the increased carbon-carbon bond distances noted in X-ray structures of coordinated olefins and acetylenes. The actual degree of back-bonding in most metal complexes is difficult to assess, but judging from indirect evidence, it appears significant enough to make a major contribution to the coordinate bond in some cases [8]. The mixing-in of the ligand's excited state through back-bonding is a significant factor in the reactivity of that ligand. It is particularly significant in the forbidden-to-allowed process where it is the central feature.

A transformation A → B is, generally speaking, symmetry-forbidden because the ground state of A corresponds to an excited state of B; and, necessarily, the ground state of B corresponds to an excited state of A. A transition metal can, through coordinate bonding, populate that excited state of A corresponding to the ground state of B. The molecular system A is thus distorted toward B upon simple coordination. If the symmetry-match of occupied molecular orbitals in the two systems is complete, A is free to transform to B with preservation of ligand-to-metal coordinate bonding. This A-to-B transformation is an example of the forbidden-to-allowed process.

We shall illustrate this process as it applies to the $[2+2]$ cycloaddition process. This treatment will be general, making no distinctions between $[\pi 2]$ and $[\sigma 2]$ systems. We assume that the ligand transformation, however, is suprafacial, i.e., $[\sigma, \pi 2_s + \sigma, \pi 2_s]$. To illustrate the role of coordinate bonding in the forbidden-to-allowed process, and its preservation, the transformation A → B will first involve a bisligand (A) transforming to a bisligand (B).

The bisligand coordinate bond is represented by four molecular orbitals, two of donor character and two back-bonding. In the perspective used in Fig. 3, the ligands rest in a plane above the metal; the Z axis points upward, between the ligands and through the plane perpendicular to it. The d orbitals were selected only to illustrate the required symmetries. A better molecular orbital representation would mix metal s, p, and d combinations of the appropriate symmetries. Mixing, however, does not alter the picture as we shall describe it; single orbitals will be used for simplicity. The ligand molecular orbital combinations in Fig. 3 are represented by simple vertical blocks with the indicated symmetries and

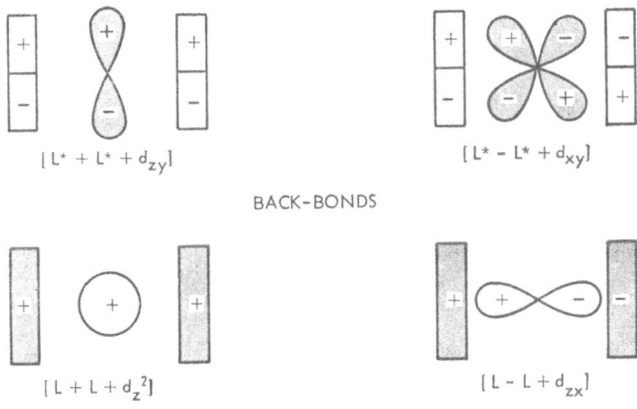

Fig. 3. Diagrammatic representation of the four molecular orbitals composing the bisligand-metal coordinate bond

shading to denote the centers of higher electron density. The ligand bonding orbitals are thus shaded and are of donor character. The antibonding ligand combinations are unshaded and are in back-bonding molecular orbitals. The actual ligand orbitals could be either olefin π and π^* or σ and σ^* or could come from a combination of a σ and π bonds (*i.e.*, $L_1 = \pi$, $L_2 = \sigma$; $L_1^* = \pi^*$, $L_2^* = \sigma^*$).

The molecular orbitals in Fig. 3 denote a bisligand system (A) interacting with a metal center in a full coordinate bonding manner. Consider now the transformation of A to B through a $[\sigma, \pi 2_s + \sigma, \pi 2_s]$ process with the metal removed. This process is easily visualized through the correlation diagram in Fig. 4. The forbiddenness of the transformation stems from the crossing of the Ψ_2 and Ψ_3 molecular orbitals in A. Symmetry conservation requires a symmetry-match of the molecular orbitals indicated (*i.e.*, $\Psi_1 - \Psi_1'$; $\Psi_2 - \Psi_3'$; $\Psi_3 - \Psi_2'$; $\Psi_4 - \Psi_4'$). The ground state occupations of systems A and B are indicated in the Figure by shading. Quite clearly, if the electrons in A were allowed an unhindered path to B (configuration interaction aside) an excited state of B (*i.e.*, Ψ_1' (2), Ψ_3' (2), Ψ_2', Ψ_4') would be generated. This ground state — excited state correlation, the crux of the symmetry-forbidden assignment, means an energy barrier to the A → B transformation, and one that can be significant, reflecting the energy separating the two states.

When we inject the metal into this system in a full coordinate bonding manner, the picture changes. Consider the correlation diagram in Fig. 5. Note first that the critical crossing of Ψ_2 and Ψ_3 orbitals still occurs. The bonding electrons in A, however, find a smooth and continuous path to

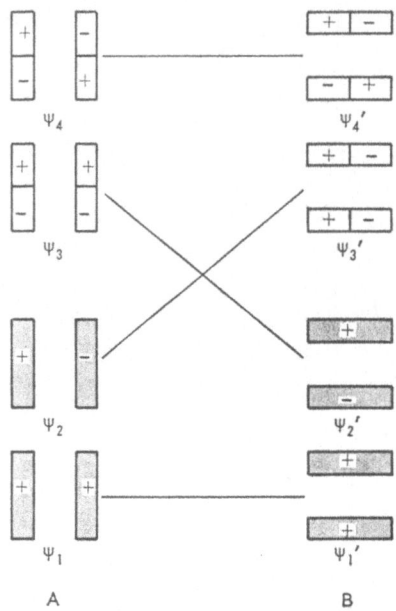

Fig. 4. Correlation diagram of bisligand system A transforming to bisligand system B with the metal removed

bonding orbitals in the ground state of B. The forbiddenness to the A → B transformation is removed by the metal which provides an empty orbital of Ψ_2 symmetry (e.g., d_{zx}) and a filled orbital of Ψ_3 symmetry (e.g., d_{zy}), thus allowing the indicated exchange of electron pairs to occur. It is important to note that both ligand systems A and B are in ground states and compare to the representation in Fig. 4; thus, both enjoy full coordinate bonding to the metal (cf. Fig. 3). The metal coordinated to B (indicated in brackets in Fig. 5) is ordered differently from that coordinated to A. The exchange of electron pairs results in a spatial redistribution of metal valence electrons, e.g., $d_{zx}, d_{zy}(2) \rightarrow d_{zx}(2), d_{zy}$. The redistribution is essential to the overall retention of coordinate bonding. However, it can create problems within the metal system in certain ligand fields, and we shall treat this subject in a later section.

We have noted earlier that back-bonding mixes a portion of the ligand's excited state into the ground state of the coordinated ligand. This distorts the ligand structurally, moving it in the direction of its populated excited state. This aspect of coordinate bonding is particularly pertinent to the forbidden-to-allowed process. Electronic population of Ψ_3 in A, for example, creates bonding in the *bonding* regions of B (Ψ_2'). The diminished population of Ψ_2 (A) from ligand-to-metal donor interaction has a similar effect in

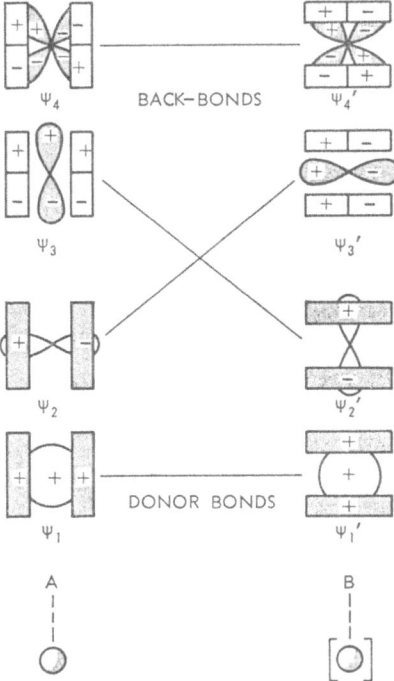

Fig. 5. Correlation diagram of bisligand system A transforming to bisligand **system B** with preservation of full coordinate bonding

distorting A toward B. Indeed, coordinated A and B have an orbital relationship to each other such that full coordination of either species alters its structure to some point between the two; the point is dictated by the relative thermodynamic stabilities of the two structures, the relative energies of the respective coordinate bonds and, as we shall show, the orbital symmetry restrictions associated with the ligand field of the nonreacting ligands. Bidentate coordinate bonding, then, is directly related to the A → B process, and, therefore, a critical factor in the forbidden-to-allowed process.

Bidentate coordination is not an essential feature to the $[_\sigma, _\pi 2_s + _\sigma, _\pi 2_s]$ A → B process. The forbidden-to-allowed function can, in theory, be achieved through the coordination of a single ligand bond to the metal. The exchange of electron pairs between metal and coordinated ligand proceeds similarly, except that the metal may use orbitals of different symmetries. The net result, however, is the same, in that the metal experiences a reordering of its valence electrons. This process is illustrated diagrammatically in Fig. 6. Only orbitals Ψ_2 and Ψ_3 are represented since these are the centers of electron exchange; the other orbitals (Ψ_1 and Ψ_4) interact with the metal, but do not play a role in the forbidden-to-allowed process.

55

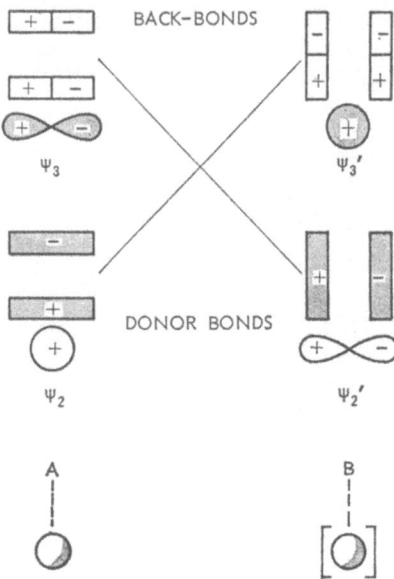

Fig. 6. Correlation diagram for the monodentate ligand system A transforming to monodentate B through the $[\sigma,\pi 2_s + \sigma,\pi 2_s]$ process

Both the bidentate and monodentate processes probably play a role in the catalysis of symmetry-restricted reactions. For some reactions, like the cyclobutanation of olefin ligands suggested in olefin metathesis [9], bidentate coordination is essential to the critical transformation. In other processes, such as the valence isomerization of quadricyclene (9) to norbornadiene (10) [10], both paths are available.

9   10

This transformation is highly energetic (the strain energy released is about 65 kcal/mole [11]) and constitutes one of the more interesting cases of metal-assisted forbidden transformations. We shall examine this case more closely by considering first the effects of bidentate coordination.

The cyclobutane ring in 9 offers two bond sets for bidentate coordination, ab and cd. The metal can accept either ab or cd positions of coordination by displaying one of the two electronic distributions indicated in Fig. 5 (i.e., A for cd and B for ab). Thus, $[d_{zy}, d_{zx}(2)]$ directs bidentate coordination to

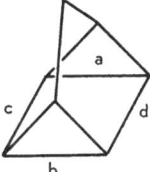

bonds ab (ligand B) and $[d_{zy}(2), d_{zx}]$ centers bidentate coordination at bonds cd (ligand A). One set of bonds should be strongly preferred over the other. If bonds ab are the centers of coordination, the excited state of 9 which mixes into the coordinated ligand is norbornadiene 10; that is, quadricyclene becomes a mixture of 9 and 10, the resonance hybrid $[9 \leftrightarrow 10]$.

If, on the other hand, bonds cd are the centers of coordination, the excited state of 9(A) is the dicyclopropenyl system 11(B). Quadricyclene thus becomes $[9 \leftrightarrow 11]$. The relative stability of 11 is not known, but the high ring strain of the cyclopropene system (ca. 52 kcal/mole) should place

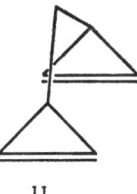

11

it below 9 and 10[11]. Bidentate coordination to quadricyclene should reasonably be focused at bonds ab, with the metal ordering its electrons as indicated in B (Fig. 5), thereby transforming the polycyclic ring system toward 10, releasing energy and relaxing ring strain in the process. Moreover, conceivably nothing prevents total relaxation to the preferred valence isomer 10, with preservation of a full metal-to-ligand coordinate bond (cf., B → A, Fig. 5).

A formally symmetry-forbidden $9 \rightarrow 10$ path has been suggested for these processes [12]. Along this path, the metal orders its valence electrons as in A (Fig. 5), thus focusing bidentate coordination at centers cd of 9. Quadricyclene then transforms to norbornadiene (10) through configuration interaction. The injection of configuration interaction here is reasonable, allowing greater flexibility through the mixing-in of additional higher excited states. It is doubtful, however, that a ligand system such as 9 will prefer coordination through bonds cd with the corresponding distortions toward 11. The forbidden-to-allowed path has the advantage of providing continuous and full metal-to-ligand bidentate coordinate bonding across the $9 \rightarrow 10$ reaction coordinate. The coordination of 9 to a metal through ab provides a symmetry match with coordinated *10*, thus creating in *9* a strong propensity to transform to *10*. Coordination of *9* through bonds *cd*, in contrast, mixes into the ground state of *9*, that of *11*, creating a propensity to transform in that direction. The energetics of these two modes of coordination would seem to be sharply different, the former being the preferred.

The reaction path from monodentate coordination of *9* is similar to that from bidentate coordination. Again, ligand system *9* can direct bonds *ab* or *cd* to the metal. Inspection of the correlation diagram in Fig. 6 shows that coordination to *a* or *b* opens a symmetry-allowed path to *10* and coordination to bonds *c* or *d* opens a path to *11*. The strongly preferred centers of coordination should be bonds *a* and *b*.

The high strain energy of quadricyclene makes it a good candidate for the monodentate path of transformation. This would, of course, be the preferred path of reaction on those metal systems where bidentate coordination is obtained only with difficulty. Further, the high strain energy of quadricyclene [*9* contains about **65 kcal/mole** more strain energy than *10*[11)]] means that it shall possess strong mono and bidentate bonding character. Coordination through the preferred centers relaxes *9* to *10* with the corresponding release in strain energy. Systems such as *9*, then, should be very special ligands carrying with them, in the form of strain energy, bonding potential which can be unlocked and released by metal centers. Quadricyclene should be a powerful ligand posessing unusually strong metal affinity. These systems are deceptively saturated. They possess unsaturated bidentate centers of coordination immediately available upon interaction with a metal center and thus should be strong competitors for bidentate sites on the coordination spheres of metals.

Where the thermodynamic driving force for the $[_\sigma, _\pi 2_s + _\sigma, _\pi 2_s]$ A → B transformation is not great, valence isomerization through bidentate coordination would seem to dominate, particularly when the ligand has good bidentate bonding properties. Ligand A should better survive monodentate coordination, assuming bidentate coordination in a stepwise manner. If the coordinate bonding properties of B are as good as (or better than) those of A, A can undergo smooth transformation to B, preserving a full bidentate coordinate bond in the process. An example of this type of transformation is the valence isomerization of *exo*-tricyclooctene *12* to the tetracyclooctane

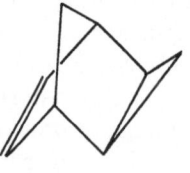

12

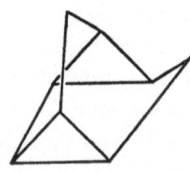

13

13[13]). This transformation constitutes a $[_\pi 2_s + _\sigma 2_s]$ process. Both *12* and *13* appear to be good bidentate ligands. The cyclopropane ring is known to coordinate edgewise to metals [14]. This is not surprising since the ring system possesses about 27 kcal/mole of strain energy [11] to supply, in part, upon coordination. Of further note is the fact that only two centers for bidentate coordination are available in *12*, unlike *9* which possesses four (*i.e.*, *ab* and *cd*). The possibility of a formally symmetry-forbidden transformation [12] to *13* is, therefore, even less likely. It would require a metal ordering its valence electrons as described in B (Fig. 5), thus focusing bidentate coordination into regions of space void of bonds. The bidentate coordination properties of *12* will more reasonably order the metal valence electrons as described in A (Fig. 5), thereby establishing a full metal-to-ligand coordinate bond and creating a propensity to transform to its preferred valence structure *13*.

The role of coordinate bonding in the forbidden-to-allowed [2+2] process can be summarized briefly. Coordination of a ligand system to a metal mixes portions of the ligand's excited state into the ground state of the coordinated species. It is in the nature of the coordinate bond to create within the ligand system a propensity to transform to a valence isomer reflecting that excited state encountered along an otherwise symmetry-forbidden path. Orbital symmetry restrictions are relaxed through metal coordination; the ligand is free to transform to a new ligand system with an exchange of electron pairs with the metal. In the forbidden-to-allowed process, the reordering of metal valence electrons is such that metal-to-ligand coordinate bonding is preserved. A bidentate ligand system, for example, is free to rearrange to a new bonding configuration rotating its centers of bidentate coordination 90° with preservation of full bidentate coordinate bonding. This process is illustrated diagrammatically in the transformation *14* ⇌ *15*. Back-bonding electron density is shifted from the

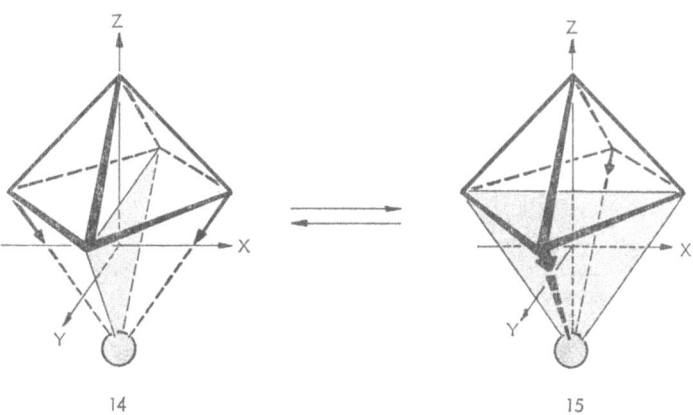

14   15

ZY plane in *14* to the ZX plane in *15*, preserving bidentate coordination at the centers indicated by arrows.

B. Stepwise Processes

A symmetry-forbidden transformation A → B can be catalyzed through a series of distinct steps, each of which being symmetry-allowed. The reaction A → B can be catalyzed along the path described in Eq. 7, through the intermediacy of the metal-bonded species X.

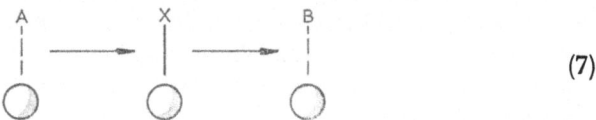

$$(7)$$

In this idealized stepwise process, the exact nature of the metal-X bond is not specified. We assume that it differs in kind from the coordinate bonds to A and B. The ligand system X, moreover, is best considered a distinctly different molecular species from either A or B. We shall continue to concern ourselves here with $[\sigma, \pi 2_s + \sigma, \pi 2_s]$ processes, although the principles discussed will be applicable to a variety of symmetry-forbidden transformations. In the catalysis of these reactions, Class 3-type reactions should play the major role in the stepwise processes. The metal, here, will serve as a participant in the transformation A → X, enmeshed in the bonding network of species X. This can be considered an oxidative insertion process (Eq. 8), for example. The overall process A → B, then, can be looked upon

$$(8)$$

as an oxidative insertion followed by a reductive elimination. The metal, however, attached to B will be ordered differently from that coordinated to A, the forbidden transformation A → B having imposed a spatial redistribution of metal valence electrons similar to that in the forbidden-to-allowed process (cf. Eq. 9). Consider, for example, the cyclobutanation of a

$$(9)$$

bisolefin ligand system *16* through the metalocyclo species *17*.

The first step (*16* → *17*), to be symmetry-allowed, must proceed along a $[_d2_s + _\pi2_s + _\pi2_s]$ path and thus withdraws from the metal an electron pair from one of its symmetric (relative to the mirror plane between the olefin ligands, *i.e.*, the ZY plane in coordinate system *6*) orbitals. The second step, reductive elimination (*17* → *18*), constitutes a symmetry-allowed $[_\sigma2_a + _\sigma2_s]$ process. The extruded cyclobutane ring must take with it an electron pair in a symmetric (relative to the ZY plane) σ bond and thus constitutes the $[_\sigma2_s]$ participant. The metal, therefore, assumes the role of the $[_\sigma2_a]$ participant and withdraws an electron pair from the metalocyclo bond system through one of its antisymmetric orbitals. In the overall process, then, the metal has passed from $[_d2_s]$ in *16* to $[_d2_a]$ in *18*. The reverse process (*18* → *16*) will behave similarly involving an overall redistribution of two metal valence electrons from an antisymmetric atomic orbital in *18* to a symmetric orbital in *16*. This is exactly the process encountered in the concerted transformation when the metal center operates from monodentate coordination. It is not surprising, therefore, that clear distinctions need not exist between what may be considered concerted and stepwise processes.

A process is clearly stepwise when species X and A are in different bonding configurations and not simply extensions of each other resting somewhere along a continuum. The metalocyclo species *17*, for example, cannot be easily confused with the bisolefin system *16*. On the other hand, a metalocyclo intermediate may not be clearly distinguishable from a simple coordinated ligand. We have already noted the similarities between a simple, edgewise-coordinated cyclopropane ligand (*7*) and the four-membered metalocyclo species *8*. A spectrum of bonding degrees undoubtedly exists between them. In viewing the process *18* → *16*, therefore, one must be careful in distinguishing the so-called stepwise process involving the distinct intermediate *17* and the concerted process proceeding through monodentate coordination. In this latter process, coordination of the σ bond of the cyclobutane (or cyclopropane for other related processes) ring results in a reduction of its bond order. The coordinated bond, through back-bonding, is thus partially broken, moving in the direction of *17*. The degree of bond order reduction will depend on ring strain and the energy and overlap properties of the metal orbitals. The coordinated ligand in *18* can then undergo transformation to *16* in a concerted manner with the partially broken bond (through coordination) and its parallel partner breaking simultaneously. The stepwise process is only clearly distinguishable from this process when the carbon-carbon bond in the cyclobutane ring is completely (or essentially that) cleaved upon coordination to the metal center generating intermediate *17*. Obviously, these two processes bear a strong resemblance to each other, suggesting a rather broad grey area separating extrema.

The question of stepwise vs. concerted processes in the catalysis of symmetry-forbidden reactions becomes more important when the postulated forbidden-to-allowed process itself is questioned. Certainly, stepwise processes in catalysis are common, and have been known to intervene in catalytic reactions for some time. So to establish that an overall chemical transformation A → B proceeds through distinct steps (*e.g.*, A → X → B) is not, in itself, chemically significant. To suggest, however, that a chemical transformation A → B can proceed smoothly on the coordination sphere of a transition metal, and essentially nowhere else, introduces the possibility of novel chemistry, not encountered in or outside of catalysis. The mere observation that a symmetry-forbidden reaction A → B proceeds in the presence of a metal, does not, in itself, mean that the symmetry-restricted transformation A → B proceeded in a concerted manner on the coordination sphere of that metal. Here, the question of stepwise vs. concerted takes on new significance, for it bears on the possible existence of a special kind of chemical transformation.

To clearly establish experimentally that a given transformation A → B proceeds in a concerted manner is difficult. The introduction of an additional factor, a species with catalytic properties, can only compound the difficulty. The situation is further complexed by the frequent fact that the actual active molecular species in a catalytic process is not itself known, cannot be isolated or identified. To set out to establish a stepwise mechanism by trapping intermediates, therefore, is a reasonable and often fruitful approach, for it rules out what would otherwise be difficult to establish. But to establish a stepwise mode of catalysis in one reaction need not reflect on the mechanisms of others, even similar in kind.

A stepwise process has been suggested for the valence isomerization of *12* to *13* [15]. In this treatment experimental evidence was presented supporting the intermediacy of a species "X" which transforms to the observed products. The intermediate, however, was not isolated, trapped, or specifically identified. Intermediate "X" could very well be the reactant *12* coordinated to the metal in a bidentate manner. The critical step in this case would be the attainment of bidentate coordination. If this is so, *12* → *13* could reasonably proceed through the concerted ligand transformation with preservation of full bidentate coordinate bonding.

A better example for the possible intervention of a distinct intermediate was recently disclosed [16] in a study of the valence isomerization of cubane (*19*) to *syn*-tricyclooctadiene (*20*) using various rhodium (I) catalysts.

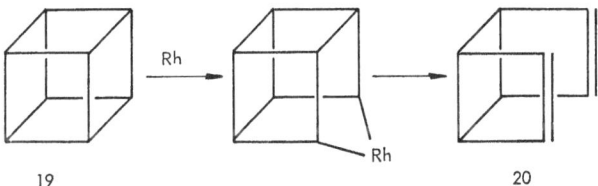

19   20

With stoichiometric amounts of $[Rh(CO)_2Cl]_2$, cubane reacted yielding an organo-rhodium compound believed to be *21*.

$$4 \ (19) \ + \ 2 \ [Rh(CO)_2Cl]_2 \ \longrightarrow$$

21

This was supported by the fact that the rhodium carbonyl species reacted with triphenylphosphene giving the cyclic ketone *22* in 90% yield.

$$21 + P(Ph\phi_3)_3 \longrightarrow$$

22

Reactions with carbomethoxy-substituted cubanes (*23*; R=COOR) were also examined. The reaction of *23* followed by triphenylphosphine addition yielded the corresponding ketones resulting from insertion into bond-type *a* and bond-type *b* in 66% and 34% selectivity, respectively. This selectivity, moreover, parallels that for the catalytic process using

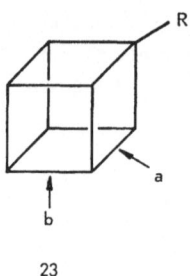

23

a variety of [Rh(Diene)$_2$Cl]$_2$ complexes (*e.g.*, Diene = norbornadiene, cyclooctadiene and *20*). Dienes *24* and *25* were formed in 29% and 71% selectivity, respectively, thus suggesting an almost statistical attack at the cubane bonds.

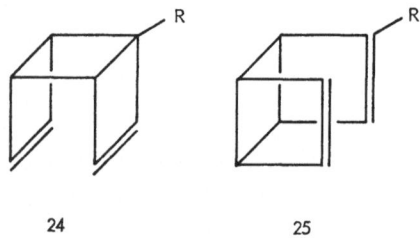

24   25

The reaction kinetics of the cubane and substituted cubane reactions were also examined with the [Rh(CO)$_2$Cl]$_2$ and [Rh(norbornadiene)$_2$Cl]$_2$ systems. The ratio of the rate constants for the two processes was found to remain essentially constant with the cubanes although the rates for the catalytic and stoichiometric processes differed by a factor of 100.

The critical step in both processes, therefore, appears the same, namely the almost statistical attack of a rhodium nucleus at cubane bonds *a* and *b*. The formation of ketone *22* is strong evidence for an oxidative insertion into a σ bond of cubane yielding a metalocyclo-species *21* as an initial step. Carbonyl insertion into the Rh—C bond would reasonably follow.

[Rh(diene)$_2$Cl]$_2$ and [Rh(CO)$_2$Cl]$_2$, however, need not react along identical reaction paths. The catalytic chemistry of highly strained cyclobutane systems can be dramatically altered by changes in the electronic character of a metal's ligands [17]. The dienerhodium-catalyzed valence isomerization of cubane to diene *20* could proceed through simple monodentate coordination and not involve a distinct metalocyclo-intermediate. The coordinated bond would thus suffer only partial cleavage, the actual degree of insertion resting anywhere along the continuum between complete carbon-carbon bond cleavage and simple coordination. Here, a clear distinction between a concerted, forbidden-to-allowed process springing from monodentate coordination and the clear stepwise process would not be possible. Since complete insertion into a carbon-carbon bond and simple coordination with carbon-carbon bond order reduction (due to back-bonding) are, to a significant degree, extensions of each other, *both would be expected to exhibit similar discriminating properties*. The self-consistent pattern in the kinetics of the rhodium chemistry of cubane, therefore, could very well be consistent with a combination of the two processes.

The role of bidentate coordination has not been discussed in this context. It can, however, be a factor in the courses of reaction open to a molecular system exposed to a transition metal complex. Consider the above system. First, the second-order rate constants for the diene catalysts ([Rh(diene)Cl]$_2$. differed significantly for different dienes indicating that the diene originally present on a catalyst remains attached throughout the catalyst's lifetime. This indicates that bidentate coordination was probably not opened to the cubane ligand, and that the seat of catalysis was centered at the single coordination position along the principal axis of the square planar system. The opportunity, therefore, for reaction via bidentate coordination, where the forbidden-to-allowed process is most favorable, was not realized in this system. Reaction apparently proceeded on a single coordination site where two reaction paths were open, namely the stepwise process through oxidative insertion to a distinct intermediate and the concerted, forbidden-to-allowed process proceeding from monodentate coordination.

The stepwise, oxidative cycloaddition mechanism [particularly with d^8 metal systems [18)] could intervene in the valence isomerizations of strained, cyclobutane ring systems where energy factors and difficulties in attaining bidentate coordination work in its favor. For the other processes, however, where bidentate coordination is either very favorable or guaranteed, its contribution to catalytic chemistry would seem to be significantly less.

C. Symmetry Restrictions to Reactions

1. The Reaction of Acetylenes

Simple olefins undergo cycloaddition, one way or another, in the metathesis process with remarkable ease; *e.g.*, with a tungsten halide catalyst at room temperature, 2-pentene is converted to an equilibrium mixture of 2-butene, 2-pentene and 3-hexene in a few seconds [9)]. One should reasonably expect a similar transformation of simple acetylenes, *e.g.*, 26 → 27. The transformation 26 → 27 might appear, at first inspection, to be symmetry-allowed. Considering the unusually attractive ligand properties of cyclobutadiene[19)], what could prevent two coordinated acetylene ligands from fusing to a cyclobutadiene ligand? Transformation 26 → 27, however, encounters

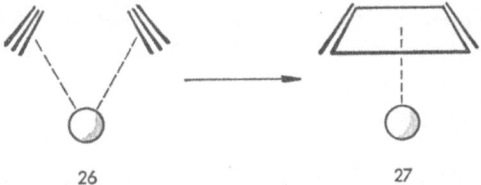

26   27

symmetry restrictions with all transition metals [20,29]. They stem from anti-bonding interaction between the nonreacting set of π bonds and the metal as cycloaddition proceeds.

Consider, for example, the reordering of metal valence electrons which attends the $[_{\pi}2_s + _{\pi}2_s]$ fusion of acetylene π bonds, *i.e.*, *28 → 29*.

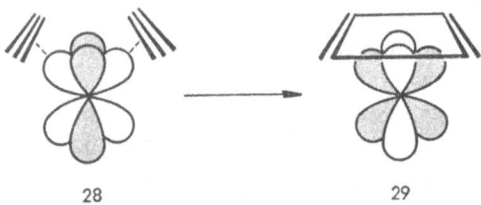

28   29

In coordinate system *6*, and assuming for simplicity pure d orbitals, this constitutes a $[d_{zx}, d_{zy}(2) \rightarrow d_{zx}(2), d_{zy}]$ transformation. It proceeds, necessarily, from the forbidden-to-allowed cycloaddition of the acetylene π bonds. This final distribution of metal valence electrons [*i.e.*, $d_{zx}(2), d_{zy}$], however, is the opposite to that required [*i.e.*, $d_{zx}, d_{zy}(2)$] for a coordinate bond to the incipient cyclobutadiene. The restrictions we are referring to here are encountered all along the reaction coordinate and have nothing to do with the final state of *29*. Certainly, a metal system would not adopt the electronic distribution implied in *29* when interacting with a cyclobutadiene ligand. The metal, however, is locked into this electron distribution by the transforming ligand system. It might further seem that *29* can just as well be considered *30* at some point along the reaction coordinate.

30

The ligand system represented by *30* is, in fact, that which combines with metal system $[d_{zx}(2), d_{zy}]$ yielding a full coordinate bond. Ligand system *30*, however, is not a legitimate, equal representation of ligand system *29* at the early stages along the reaction coordinate. The $(\pi - \pi)$ combination of bonds in *30* corresponds in symmetry to the $(\pi^* + \pi^*)$ combination of orbitals in *29*. There is, similarly, a symmetry match between the $(\pi - \pi)$ combination in *29* and the $(\pi^* + \pi^*)$ combination in *30*. Ligand representation *29*, then, bears a symmetry-forbidden $[_\pi 2_s + _\pi 2_s]$ relationship to ligand system *30* at the early stages along the reaction coordinate. When the cyclobutadiene ligand reaches a point along the reaction coordinate where it possesses a four-fold axis of symmetry, the $(\pi - \pi)$ and $(\pi^* + \pi^*)$ combinations in *29* become energetically equivalent, and, ignoring the metal for the moment, *29* can be replaced by *30*, providing we have allowed for the appropriate redistribution of ligand electron pairs [*i.e.*, from $(\pi - \pi)$ in *29* to $(\pi - \pi)$ in *30*]. But no easy way seems to exist for electron manipulation of this kind early along the reaction coordinate, and, consequently, the incipient cyclobutadiene ligand should be frozen to bonding configuration *29* which interacts in a negative way with d electron configuration $[d_{zx}(2), d_{zy}]$.

It is instructive to depart here from the general subject and discuss briefly the bonding between a metal center and a cyclobutadiene ligand.

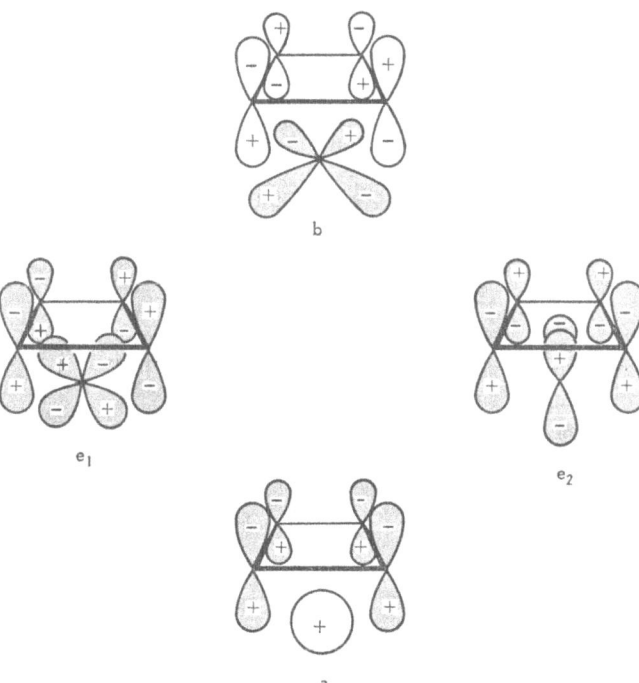

Fig. 7. The four bonding molecular orbitals in the cyclobutadiene (C_{4v})-metal complex

To be consistent, we shall adopt the simple orbital representations used above to describe coordinate bonding interactions. Cyclobutadiene, being a bidentate ligand system, is described by four bonding molecular orbitals. They are similar in symmetry to the general bidentate system described earlier (cf. Fig. 5) and are represented in Fig. 7.

Of the four orbitals, the a is clearly of donor character and the b, back-bonding. The two e orbitals, however, are degenerate, and clear donor or acceptor assignments cannot be made. They can be made arbitrarily by making one of the ligand combinations (e.g., e_1) a $(\pi - \pi)$ and the other (e_2) a $(\pi^* + \pi^*)$. The d_{zx} in this example would be unoccupied in e_1 and the d_{zy} would be occupied in e_2; these molecular orbitals would then be of donor and back-bonding character, respectively. However, the opposite assignments are just as appropriate.

These two possibilities can be represented a different way. The $(\pi - \pi)e_1$, $(\pi^* + \pi^*)e_2$ assignments are represented by 31 and $(\pi - \pi)e_2$, $(\pi^* + \pi^*)e_1$ by 30.

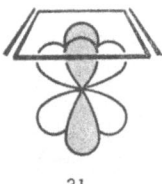

31

Structures 30 and 31, taken together, constitute a good representation of the special coordinate bond of the cyclobutadiene ligand, $i.e.$,

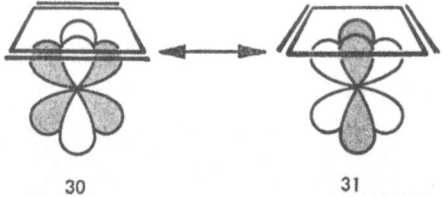

30   31

Return now to transformation $28 \rightarrow 29$. 28, as it traverses the reaction coordinate to 29, clearly does not represent a bonding state of the cyclobutadiene metal complex. In fact, 29 represents the situation where both e_1 and its antibonding counterpart (not represented in Fig. 7) are occupied and e_2 is empty. The ligand-to-metal bonding, then, associated with two important molecular orbitals in the cyclobutadiene ligand system (e_1 and e_2) is not realized with π bond fusion of a bisacetylene ligand system.

The symmetry restrictions discussed here can be removed through the intervention of a second metal interacting at the opposite face of the incipient cyclobutadiene (at the apices of the C_{2h} complex 32).

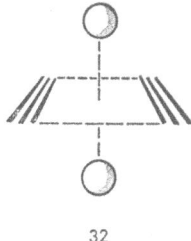

32

The function of the second metal is similar to that of a single metal operating on one set of π bonds. Very simply, it operates on the remaining π bonds in *29* in a forbidden-to-allowed manner, transforming that system to *30*, which enjoys a full bidentate coordinate bond to both metal functions. This process can be represented by the transformation *33 → 34*.

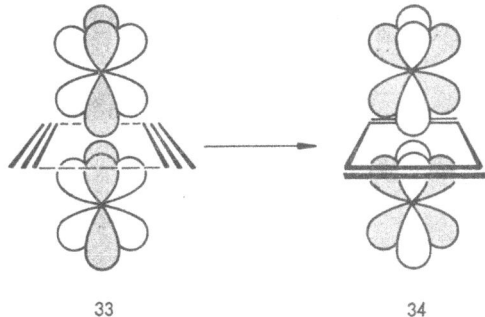

33   34

The ligand-to-metal bonding represented by *34*, of course, differs from that represented by *30*. In *34*, the cyclobutadiene ligand is shared by two metal centers, thus reflecting the bonding properties of a three-centered bond. There will, therefore, be an extra set of orbitals corresponding to those in Fig. 7, which are nonbonding in character.

The concepts discussed regarding the symmetry restrictions and their removal can be described in a number of ways. Complete correlation diagrams can be constructed and the forbiddenness illustrated by sharp orbital crossings [20]. Although definitive, this approach would not as clearly illustrate the nature of the restraints to reaction.

Whe can briefly summarize the case for acetylenes. A single metal can, in an allowed manner, fuse two acetylene π bonds to two σ bonds; this is represented by *28 → 29*. But *29* is not a representation of a cyclobutadiene-metal coordinate bond; *30 ↔ 31* is. Full cyclobutadiene-metal bonding can be generated across the reaction coordinate by introducing a second metal (*32*) that operates on the second set of π bonds as the first does, thus giving *33 → 34*.

Cyclobutadiene-metal complexes were obtained from reactions of the corresponding acetylenes with metal complexes [19b,21]. Orbital symmetry principles would suggest that these complexes are either formed via step-wise processes or involved the intervention of bimetallic species (32). The stepwise routes are particularly attractive. Acetylene ligands can reasonably be expected to undergo a $[_d2_s + _\pi2_s + _\pi2_s]$ cycloaddition with the metal center generating the metalocyclodiene intermediate 35. Cyclobutadiene can then be extruded from the metal center with the aid of another metal.

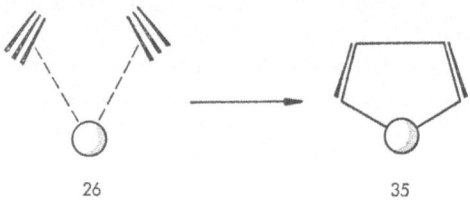

26   35

Indeed, a spectrum of bimetallic species containing the 35 moiety was observed [19b], e.g., 36. The involvement of two metal centers in a forbidden-

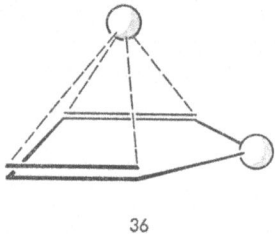

36

to-allowed way is also suggested for some systems. $[Co(CO)_4]_2Hg$ reacts with substituted acetylenes (90—100 °C) to yield the polymetallic complex believed to have structure 37 [22].

$(CO)_4Co$——Hg——Co⋯⋯Co——Hg——$Co(CO)_4$

37

Compound *37*, of course, need not reflect its origin. It could just as well have resulted from intermediates like *36* as *32*. What is significant about *37* is that some metal systems apparently prefer face-sharing coordination to a cyclobutadiene ligand. A similar bonding configuration (like *32*) might be anticipated for acetylene ligands. Bonding configuration *32* could thus be more than a theoretical curiosity, and may actually intervene in the reactions of acetylenes with certain metal systems.

Acetylenes are well known to undergo facile trimerizations to derivatives of benzene in the presence of various transition metal catalysts [23]. A number of mechanisms for this process have been considered including the intervention of metal-cyclobutadiene complexes [24]. This chemistry, however, was subjected to close examination by Whitesides and Ehmann, who found no evidence for species with cyclobutadiene symmetry [25]. Cyclotrimerization of 2-butyne-1,1,1-d_3 was studied using chromium(III), cobalt(II), cobalt(0), nickel(0), and titanium complexes. The absence of 1,2,3-trimethyl-4,5,6-tri(methyl-d_3) benzene in the benzene products ruled out the intermediacy of cyclobutadiene-metal complexes in the formation of the benzene derivatives. The unusual stability of cyclobutadiene-metal complexes, however, makes them dubious candidates for intermediates in this chemistry. Once formed, it is doubtful that they would undergo sufficiently facile cycloaddition with acetylenes to constitute intermediates along a catalytic route to trimers.

The transition metal-catalyzed chemistry of acetylenes appears consistent with the symmetry principles described in this section. Much of the catalytic chemistry of acetylene very likely proceeds on a single metal center. In this case, smooth transformation to a cyclobutadiene ligand (*i.e.*, *26* → *27*) should be a rare event. Cyclobutadiene intermediates are clearly not involved in the oligomerization of acetylenes involving a variety of metal complexes. Considering the remarkably facile interconversion of simple olefins found in olefin metathesis and the known stability of the cyclobutadiene-metal system, the failure of simple acetylenes to undergo a similar and at least equally facile transformation seems best explained through symmetry restraints.

Acetylenes have been reported to undergo a metathesis-like transformation [26]. 2-Pentyne reacts over a heterogeneous catalyst of tungsten oxide on silica gel between 200° and 400 °C to yield 2-butyne and 3-hexyne. At 350 °C 2-pentyne (in cyclohexane solvent) undergoes 44% conversion to form an essentially equimolar mixture of 2-butyne and 3-hexyne in 53% selectivity. The *C*-4 and *C*-6 alkynes appear to be primary products, thus suggesting the intermediacy of cyclobutadienes. The reaction conditions, however, are sufficiently severe (200°—400 °C) to accommodate a variety of mechanisms leading to the indicated products, including *26* → *27*, *26* → *35*, and others involving ionic intermediates in stepwise processes [27].

2. Symmetry Restrictions Associated with Nonreacting Ligands

Orbital symmetry restrictions associated with the nonreacting set of π bonds of a bisacetylene system were just discussed. The central feature to the energy barriers was the redistribution of metal valence electrons which necessarily proceeds with the forbidden-to-allowed process. We address in this section a similar interaction, but this time associated with the metal's nonreacting ligands. In the forbidden-to-allowed process, metal valence electrons are removed from one set of metal atomic orbitals and injected into another. This means that metal valence electron density is redistributed spatially as reaction proceeds along the reaction coordinate. The actual redistribution will depend on the metal itself (the number of d electrons and its ligand system) and the ligand transformation. Whatever the situation, the new distribution of valence electron density can interact with the metal's nonreacting ligands in an antibonding way (*i.e.*, similar to that for acetylenes) creating energy barriers to reaction.

We shall address this question in its simplest approximation, assuming that the nonreacting ligands are point charges distributed in various geometries over the surface of the metal's coordination sphere. Three possible situations exist for interaction of the nonreacting ligands with the redistributed electron density:

1. The interaction is negative, generating metal-ligand (nonreacting) antibonding character.
2. No change in metal-ligand bonding character occurs with reaction.
3. The ligand interaction with the new distribution of valence electron density is positive, relative to that with the original electron distribution.

We shall consider each of these cases as they apply to the $[\sigma, \pi2_s + \sigma, \pi2_s]$ cycloaddition process. Our examples will involve bidentate ligand systems and, therefore, the exchange of electron pairs between reacting ligands and the metals will proceed through dp orbital combinations. The similar situation for monodentate coordination will be discussed briefly at the end of this section. For simplicity, the actual orbitals will not be specified, only their symmetries and spatial configurations relative to the set of nonreacting ligands will be indicated in the figures.

Consider first the redistribution of metal valence electrons that occurs with the $[\pi2_s + \pi2_s]$ cycloaddition $38 \to 39$. In the figures, d electron density

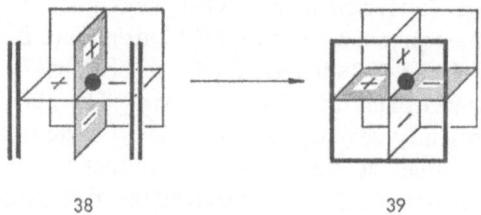

38   39

is represented by the shaded rectangles. The vertical plane contains the d_{zy}, p_y combination and the horizontal d_{xy}, p_x (coordinate system 6). The distribution of metal valence electron density in 38 focuses bidentate ligand bonding at the centers occupied by the olefin ligands. In 39, metal valence electrons have been shifted to the horizontal plane, focusing bidentate coordinate bonding at the corners of the vertical plane. This transformation represents the situation for the naked metal, and, of course, ligand field restrictions to 38 → 39 are nonexistent.

Type I. We shall consider first the case where ligand field restrictions accompany the $[_\pi 2_s + _\pi 2_s]$ process. An example is the transformation of the square planar bisolefin complex 40 to 41.

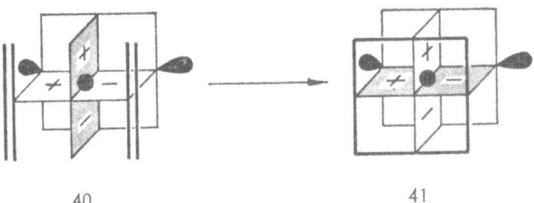

40   41

A similar ligand field restriction is encountered from the octahedral complex, although it would not be as clearly visible with the shaded figures used here. The transformation of a bisolefin system to a cyclobutane from either the square planar complex 40 or an octahedral complex will encounter energy barriers associated with the growing electron density in the XY plane (the d_{xy}, p_x orbital combination). These energy barriers will reflect the energy split dividing the two critical d orbitals (*i.e.*, d_{zy} and d_{xy}) and are associated with the strength of the ligand field defined by the non-reacting ligands.

The restrictions associated with four-coordinate complexes are reversed when the band of d orbitals is filled with metal valence electrons (*e.g.*, d^{10} systems). In these situations, ligand field restrictions are encountered from the tetrahedral complex and not the square planar. This departure from the qualitative picture based on pure d orbitals is primarily due to hybridization factors in these systems. The square planar complex requires an empty d orbital (in the plane) to construct the four hybrid orbitals in the square plane. The [2 + 2] transformation from the square planar complex thus "returns" two valence electrons (formally from a p orbital) to this orbital generating a filled d band in the process. The process proceeds without an orbital crossing. The tetrahedral system, in contrast, starts with a filled d band. The [2 + 2] process formally moves a pair of d electrons into a p orbital. This process thus involves an orbital crossing and therefore encounters ligand field restrictions.

The distribution of valence electron density in *41* need not actually result from the ligand transformation noted here. Paths to other energetically more favorable distributions exist (through configuration interaction, for example, cf. Ref. [1]). Moreover, these barriers may not be large and thus would not impose significant restraints on the ligand transformation. They are, therefore, best considered energy barriers stemming from the build-up of metal-ligand antibonding character early along the reaction coordinate and not cases of symmetry-forbidden restraints. It is important to note that the ligand transformation itself, viewed separately, has been rendered symmetry-allowed; the symmetry restrictions having been transferred to the metal complex.

For energetically marginally favorable ligand transformations such as cyclobutanation of simple olefins (ethylene, propylene, etc.), symmetry restrictions of Type I would seem to preclude reaction. The ordering of metal valence electrons in *40* focuses bidentate coordinate bonding at the olefin position, thus creating in the bisolefin system a propensity to transform to the cyclobutane (cf. Fig. 5). Molecular transformation along that mode, however, is resisted by the negative electronic interaction with the *trans* ligands. The situation should be further restricted by the changes in ligand-to-metal coordinate bonding associated with *40 → 41*. The bisolefin system would seem a better ligand system than the cyclobutane ring. The ligand transformation, therefore, gaining very little thermodynamically, might, in addition, lose coordinate bonding energy.

With ligand transformations that are energetically favorable and where bidentate coordinate bonding is strong in both valence isomers, the situation is quite different. We shall discuss this case at the end of this section.

Type II. In this category, ligand field restrictions are nonexistent. Consider the example *42 → 43*. In the trigonal prismatic complex *42* the two

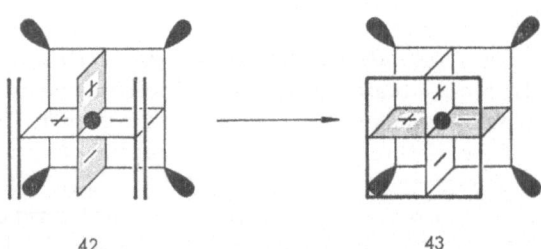

42   43

critical orbital combinations (d_{zy}, p_z and d_{xy}, p_x) are left degenerate by the nonreacting ligands. The ligand system is thus free to transform from one bonding configuration to another without altering the metal-to-ligand (nonreacting) bonding character. Other Type II systems exist, including some possessing odd coordination numbers. An example is *44*, in which the two olefins in the bisolefin system share an axial coordination position of an octahedron.

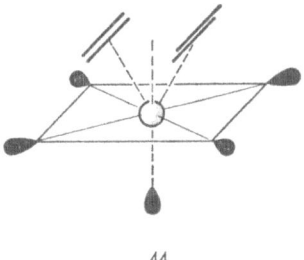

44

Intermediates like *44* are attractive candidates for the active agents in olefin metathesis. The monocapped trigonal prism (*44*) with C_{2v} idealized symmetry is not an unreasonable configuration in the very labile seven-atom family [28]. Seven-coordination, moreover, might be expected to intervene in the coordination chemistry of transition elements possessing between two and four valence electrons. An intermediate like *44* could be formed from octahedral complexes in which the axial position of coordination is occupied by the first olefin reactant. That coordination site is then shared by the two reactants in *44*. Metal systems, therefore, capable of facile seven-coordination to complexes with C_{2v} symmetry and possessing the critical number of valence electrons should exhibit pronounced catalytic activity in the catalysis of $[2_s + 2_s]$ molecular transformations. Other common ligand (nonreacting) geometries describing nonrestrictive ligand fields for the $[2+2]$ process are described elsewhere [29].

Type III. In this category of reaction, metal valence electrons are moved into energetically preferred spatial configurations with respect to the non-reacting ligands. An example is the transformation of the trigonal bipyramid complex *45* to *46*. This process is similar to Type II transformations in that

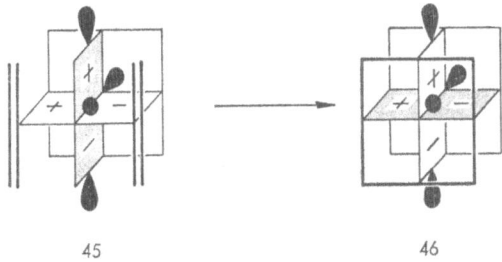

45   46

restrictive ligand fields are not encountered. Moreover, the localization of metal valence electrons into a spatially preferred configuration may provide driving force for the $[2_s + 2_s]$ transformation. There is one important difference, however. In the forbidden-to-allowed process, the centers of bidentate coordination are rotated 90°. With Type III reactions, a new geometry necessarily attends the $[2_s + 2_s]$ transformation. The implications here are not entirely clear. This process is somewhat similar to the intramolecular rearrangements of polytopal isomers. The dynamics of the two processes may bear similarities; *e.g.*, tetrahedral and octahedral complexes generally have higher barriers to polytopal rearrangement than five-coordinate complexes [28]. The structural rearrangement aspects of Type III processes are, of course, also found in Type I reactions, which, in addition, suffer from restrictive ligand field effects. Type II reactions, in contrast, are free of ligand field restrictions and remain geometrically undisturbed with respect to the rotation of bidentate coordination positions. In the transformation *42 → 43*, the bidentate centers of coordination in *42* are at the corners of the horizontal plane; the six ligands thus define a trigonal prism. In *43*, the coordination centers have rotated 90° to the corners of the vertical plane. This display of coordination sites describes another trigonal prism (*i.e.*, the indicated ordering of valence electrons in *43* points bidentate coordination toward the horizontal bonds of the cyclobutane ring). The same general retention of coordination geometry would apply to a $[2+2]$ transformation of *44*. This feature of Type II systems makes them unusually attractive as catalysts for valence isomerizations in which valence isomers are good bidentate ligand systems (*e.g.*, *9 → 10* and *12 → 13*).

We have thus far discussed rather generally the nature of restrictive ligand fields. We noted that these restrictions, even when small, could be sufficient to block molecular transformations which are, by themselves, thermodynamically only marginally favorable. Thus, simple olefins coordinated to metals that prefer square planar (with d^8 and lower metal complexes) and octahedral coordination should not be expected to undergo $[2_s + 2_s]$ cycloaddition. For molecular transformations which are energetically more favorable, the situation is different. Here, the energy provided by the exothermic $[2_s + 2_s]$ process can reasonably be expected to compensate for attending restrictive field effects. We shall now consider the valence isomerization of quadricyclene (*9*) to norbornadiene (*10*).

Quadricyclene undergoes facile valence isomerization to norbornadiene in the presence of a variety of metal complexes, including rhodium(I)

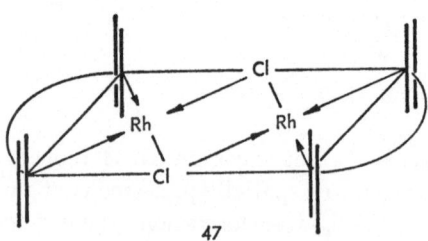

47

systems [10]. Catalysis with di-μ-chloro-*bis*(norbornadiene)dirhodium(I) (*47*) is particularly pertinent because this rhodium halide system is believed to prefer square planar coordination [30], thus placing it in the restrictive, Type I category. We have already discussed the bidentate coordination character of the quadricyclene ring system. To recapitulate, *9* should be a strong bidentate ligand with the preferred centers of coordination fixed at the cyclopropane σ bonds. Metal coordination to these centers distorts quadricyclene's molecular framework, moving it in the direction of norbornadiene. The coordination-induced transformation *9 → 10* proceeds with metal-ligand retention of bidentate coordination. Since *10* possesses about 65 kcal/mole less strain energy than *9*, this molecular distortion would constitute a relaxation and should proceed exothermically, the amount of energy released being proportional to the degree of coordination.

In this view, labile ligands attached to a metal should be highly vulnerable to displacement by *9*, particularly if bidentate sites of coordination are not fully shielded. If the bonding corridors to bidentate coordination extend sufficiently beyond the primary coordination sphere, even very weak attachment to the extended sites could conceivably launch quadricyclene on a downhill path to full coordination with displacement of the labile ligands. At any particular point along this path, the process could be reversed *only* if the strain energy required to transform coordinated *9* back to its original bonding configuration is returned to the distorted ligand. There should be a point of no return beyond which *9* is locked along a cascading energy track to *10*.

In considering the restrictive ligand field effects associated with *47*, we shall assume that a low energy path to bidentate coordination exists. *47*, of course, has open coordination positions from which reaction (*9 → 10*) can proceed via monodentate coordination (either stepwise or concerted). But we are not now concerned with the actual mechanism of valence isomerization. Complex *47* is simply used as a convenient model to illustrate the possible consequences of a high energy valence isomerization proceeding on a metal possessing a restrictive ligand field.

Given a corridor of entry, *9* should adopt a square planar configuration with *47*. The process of achieving full coordination transforms *9* to *10* and

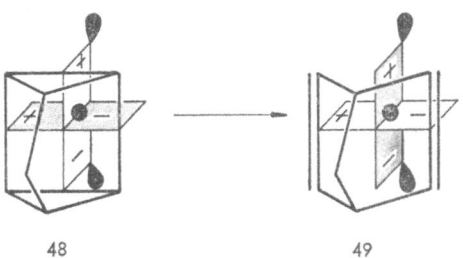

48   49

the geometry of the final complex is tetrahedral, as illustrated in *48 → 49*. This valence isomerization puts metal valence electrons into the plane containing nonreacting ligands, thus generating antibonding character and a barrier to valence isomerization. These factors, however, should not block complete molecular transformation.

The rhodium(I) halide system *47* is known to be quite labile, capable of assuming coordination numbers from three to five [30]. The lower coordination state very likely stems from the bonding character of the bridging halogen ligands. Only one halogen is fully bonded to each rhodium center (formally); the other acts in a donor capacity. Halogen lability, therefore, should be anticipated. Ligand transformation *48 → 49* could easily promote a halogen rearrangement, such as *50 → 51*.

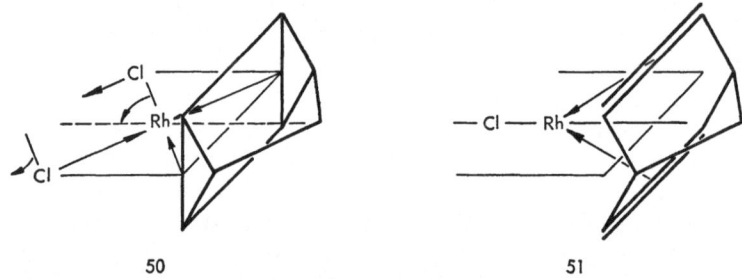

50   51

The energy generated thus drives a donor halogen ligand off the primary coordination sphere, thereby allowing the σ-bonded halogen to assume a position along the twofold axis of symmetry. Geometry *51* is a Type II system; the halogen in that position leaves the critical d orbitals degenerate, allowing essentially free electron redistribution.

Alternatively, two nonreacting ligands could rotate 90° as valence isomerization *9 → 10* proceeds, *e.g.*,

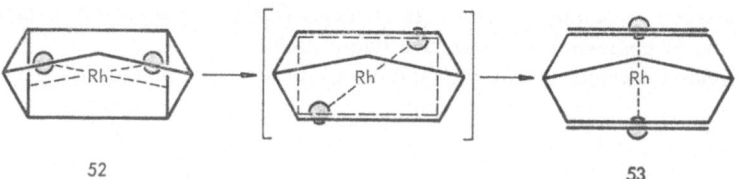

52   53

The redistribution of metal valence electrons created by the valence isomerization would support this molecular rearrangement.

Energy barriers due to restrictive ligand fields can, in theory, be circumvented in other ways. Perhaps the most interesting involves the use of photons to generate nonrestrictive, Type II systems from restrictive Type I systems. Photodimerization of norbornadiene to cyclobutane products assisted by metal complexes is knwon [31]. Jennings and Hill recently showed that norbornadienechromium tetracarbonyl and three cyclobutane dimers (*exo-trans-exo; endo-trans-exo; endo-trans-endo*) are products from the photolysis of $Cr(CO)_6$ in norbornadiene [32]. Irradiation of pure norbornadiene, free of metal, gave no dimer products. The diene chromium complex is believed to be a direct product of photochemical substitution of $Cr(CO)_6$, a known process [33]. Since neither norbornadienechromium tetracarbonyl nor $Cr(CO)_6$ produced dimeric products on thermal reaction with norbornadiene, the authors concluded that at least two photoprocesses were proceeding in the system:

1. Formation of norbornadienechromium tetracarbonyl from $Cr(CO)_6$ and

2. excitation of norbornadienechromium tetracarbonyl for dimerization.

The photolytic generation of an active Type II species from norbornadienechromium tetracarbonyl is an interesting possibility in this system, for example, *bis*(norbornadiene)chromium tricarbonyl.

V. Other Reactions

In this chapter we have attempted to describe rather broad principles regarding the catalysis of symmetry-forbidden reactions. Attention was perhaps narrowly focused on one type of restricted transformation, namely $[2_s + 2_s]$ cycloaddition. This choice of reaction was somewhat arbitrary, but not entirely. The chemistry associated with these reactions appears at this time much broader and more interesting than what has appeared from the catalysis of other types of symmetry-forbidden reactions. There are, however, conspicuous exceptions which will be discussed briefly here.

One of the first and perhaps most interesting examples of a metal-assisted, symmetry-forbidden reaction was Reppe's synthesis of cyclooctatetraene from acetylene [34]. In a careful study of this system, Schrauzer proposed a concerted mechanism in which the four σ bonds of the cyclooctatetraene are essentially formed simultaneously [35]. He proposed an octahedral complex (54) with four acetylene ligands fitted to adjacent ligand coordination positions, spatially defining the incipient cyclooctatetraene.

A forbidden-to-allowed mechanism has been proposed which involves an exchange of electron pairs between the transforming ligands and the metal [3]. This $[2_s + 2_s + 2_s + 2_s]$ process is similar, in this respect, to the $[2_s + 2_s]$ process. As might be anticipated, there is also an attending loss in coordinate bonding due to negative orbital interactions between the non-reacting acetylene π bonds and the new distribution of metal valence electrons. These factors, however, need not play as restrictive a role in the $[2_s + 2_s + 2_s + 2_s]$ process as in $[2_s + 2_s]$. First, the thermodynamic driving

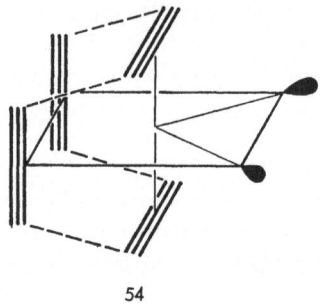

54

force should be much greater in tetramerization than in dimerization. The additional energy could compensate for the energy surrendered in the lost coordinate bonding. Further, much of the anticipated driving force for the $[2_s + 2_s]$ process was associated with the unusually good ligand properties of the cyclobutadiene ligand which are simply not achievable in the $[2_s + 2_s]$ acetylene process. Thus, this ligand transformation must essentially proceed alone, paying the energy costs for lost coordinate bonding with the energy gained from transforming two π bonds to two σ bonds. Both processes, though, lose coordinate bonding upon cyclooligomerization, the difference being that the tetramerization would appear the better able to afford it. Tetramerization differs in one other respect. With dimerization, the acetylene π orbitals maintain the same directional character with respect to the metal center as reaction proceeds; that is, the nonreacting π orbitals remain directed parallel to the two-fold axis of symmetry. The nonreacting π bonds in 54, in contrast, undergo considerable shifting and rotation along the reaction coordinate to cyclooctatetraene. This further complicates the coordinate bonding picture, since the incipient π bonds of cyclooctatetraene would describe a distinctly different spatial configuration from that implied in 54. The only conclusions that can be drawn from a simple qualitative treatment are that the symmetry restrictions to ligand cycloaddition can be removed by the metal and that some coordinate bonding would be lost upon cyclooligomerization. Consideration of the relative extent of this loss would better await a fuller treatment with molecular orbital calculations.

Other mechanisms for the Reppe process have been considered. One of the first involved the intermediacy of cyclobutadienes [19a]. This process has already been discussed in this chapter. It would appear most unlikely from theoretical principles as well as experimental evidence [25,36]. The general mechanistic aspects of acetylene oligomerization have been thoroughly reviewed elsewhere [37].

Schrauzer et al. reported a second $[2 + 2 + 2 + 2]$ cycloaddition process involving the dimerization of norbornadiene to "Binor-S" (55) [38].

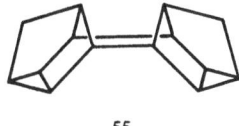

55

This remarkable catalytic reaction proceeds on a bimetallic catalyst (*e.g.*, Zn[Co(CO)₄]₂ and is believed to involve a concerted "head-to-toe" dimerization step incorporating the assistance of two metal centers (*56*). Since this reaction proceeds with high stereoselectivity, the authors proposed the π complex multicenter process in *56*. The inherent forbiddenness

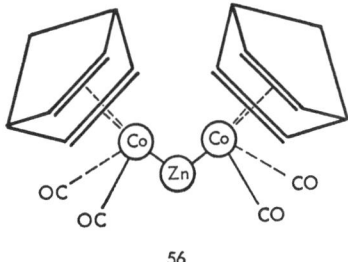

56

to the $[2_s + 2_s + 2_s + 2_s]$ process can be removed either through both metal centers as suggested in *56* or through one Co center in analogy to the $[2_s + 2_s]$ process operating through monodentate coordination. The multinuclear nature of the metal systems [39], however, would favor binuclear involvement.

A stepwise mechanism for the formation of Binor-S was offered by Katz, a strong proponent for nonconcerted processes [40]. Katz based his position on doubts regarding the actual structure of Binor-S stemming from the apparent absence of some *ir* bands in the Raman spectrum. He therefore suggested the "anti" structure *57*. Such a structure would, of course, preclude the more symmetrical involvement of metal centers proposed by

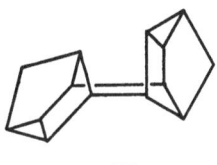

57

Schrauzer. Katz proposed instead a free radical-like mechanism in which molecular inversion through rotation about carbon-carbon bonds was possible. However, the missing bands in the Raman spectrum were subsequently located by using high resolution Raman spectroscopy [39]. The four bands are apparently weak and all close to intense Raman absorptions. The complete spectrum of Binor-S would be incompatible with structures like 57 which contain inversion centers.

Boer, Tsai and Flynn recently reported the preparation and reactions of some analogues of Schrauzer's Binor-S catalyst 56 (41). In their model system, the Zn in 56 is replaced by SnX_2 (X = Cl, Br, I, C_6H_5, CH_3). These cobalt-tin complexes exhibited catalytic activity for the dimerization of norbornadiene. Significantly, the stereospecificity of the dimerization process was found markedly sensitive to the substituent X on tin. At 60 °C, the $SnCl_2$-containing catalyst gave exclusively Binor-S, while the $Sn(C_6H_5)_2$ gave cyclobutane dimers. Dark red crystals isolated from the residues of both reactions were catalytically active, and thus were apparently intermediates in the catalytic process. The crystalline compounds proved (X-ray) to be diene complexes $[Co_2SnX_2(CO)_4(norbornadiene)_2]$ like 56, but with the dienes on the cobalt atoms oriented in a configuration approximating the transition state for the formation of cyclobutane dimers. The bond angles at cobalt, describing distorted trigonal bipyramids, were essentially the same for the two compounds. There were, however, significant differences in the two structures. In the $SnCl_2$ complex the Co—Sn—Co bond angle was 128.3° vs. 118.3° in the $Sn(C_6H_5)_2$ complex; and the Sn—Co bond was 2.50 Å vs. 2.57 Å for $Sn(C_6H_5)_2$. These differences were not attributed to steric factors, but were reasonably assigned to electronic effects by the substituents X on the Co—Sn—Co bond system.

The authors noted that the cobalt moieties in the configuration shown in the crystal structures do not align the dienes for the formation of Binor-S even when allowing for free rotation about the Sn—Co bonds. If it is assumed, however, that in the transition state both dienes are equitorial on cobalt (in a trigonal bipyramid configuration) with respect to the Sn—Co axis, an appropriate geometry for a concerted path to Binor-S is obtained. The authors argued that electronegative substituents such as Cl may weaken Co-olefins π bonds, thus labilizing the axial-equitorial dienes to rearrangement to the pure equitorial geometry. They presented nmr data supporting this proposal. At room temperature only one olefin proton signal is obtained for both the $SnCl_2$ and $Sn(C_6H_5)_2$ complexes (τ 5.82 and 6.16, respectively). The singlet splits into two equal intensity peaks at − 95° for the $SnCl_2$ compound and at − 30° for the $Sn(C_6H_5)_2$ compound. The authors attributed the coalescence phenomenon to an exchange of the diene double bonds between axial and equitorial sites of the cobalt trigonal bipyramid, which proceed, at least formally, through a pseudorotation mechanism. The fact that the coalescence temperature decreases with increasing electronegative substituents on Sn is consistent with the hypothesis that electronegative substituents X enhance diene lability opening rearrangement paths to the required geometry for Binor-S formation.

This paper also reported an X-ray structure of a derivative of Binor-S consistent with the structure originally proposed by Schrauzer (*i.e.*, *55*). There can be little doubt at this time that Binor-S is in fact *55*. Given this structure and the structural and chemical properties of the Co—Sn—Co catalytic system reported by Boer, Tsai and Flynn, and it is difficult indeed to envisage a mechanism for the formation of Binor-S other than that proposed by Schrauzer.

The concerted dimerization of norbornadiene to Binor-S is, of course, a symmetry-forbidden process. The two cobalt centers, coupled through the interacting diene ligands, can easily remove the symmetry restrictions via an exchange of metal and ligand electron pairs proceeding through coupled d (and p) orbitals in a symmetry pattern very much like that described for the $[2_s + 2_s]$ process. More important, the trigonal bipyramid structure proposed for the transition state [41], with the diene ligands resting on equitorial sites, places this system formally in the Type III category, and ligand field restrictions to the concerted cycloaddition should not be significant factors. That it appears to proceed smoothly on the coordination spheres of transition metal centers possessing nonrestrictive ligand fields lends support, to the proposal that metals can in fact remove symmetry restrictions to otherwise frobidden transformations.

Another striking example of metal-assisted symmetry-forbidden valence isomerizations involves the silver ion catalyzed rearrangement of homocubyl systems [42] (*i.e.*, *58* → *59*). A similar rearrangement was reported for cubane itself using silver salts [16]. Interestingly, other metals [*e.g.*, rho-

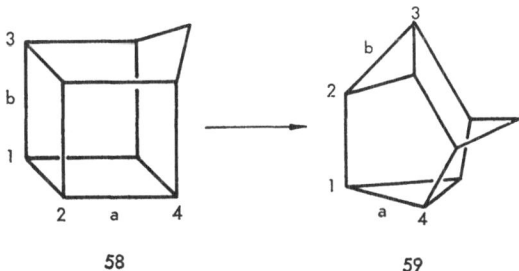

58   59

dium(I) [16]] effect an entirely different mode of rearrangement on cubane; a number of other catalyst systems (*e.g.*, mineral acid, mercuric, cuprous, zinc, iron, and certain rhodium compounds), moreover, have virtually no effect on the homocubane system [42a].

Since the rearrangement described by *58* → *59* appears to be a symmetry-forbidden $[_\sigma 2_a + _\sigma 2_a]$ process [42b], the catalytic route opened by silver is of interest. The $[_\sigma 2_a + _\sigma 2_a]$ transformation can be described by following the indicated movement of bonds ab in 58. Bond a migrates to carbon-1 and bond b moves to carbon-2. Since this process proceeds via an antarafacial mode, inversion of configuration occurs at carbons 1 and 2, while configuration is retained at carbons 3 and 4.

A catalytic function can conceivably intervene through simple metal coordination to bonds a or b (these bonds are not unique, of course, in 58) or through insertion into one of these bonds. One thing is clear, however, the spatial configuration of the migrating bonds precludes bidentate involvement with any metal center. Thus, the symmetrical kind of forbidden-to-allowed process with bidentate ligand-metal interaction suggested for catalyzing $[2_s + 2_s]$ processes is not possible here. In the catalytic process operating through one bond (either a or b), the metal can be considered a $[_d 2_s]$ participant in the process transforming it to an allowed $[_\sigma 2_s + _\sigma 2_a + _\sigma 2_a]$. The metal's role may be more clearly visualized, however, as an exchange of electron pairs where the metal, using an antisymmetric orbital, donates an electron pair to $\Psi_2(\sigma^*)$ and withdraws an electron pair from $\Psi_1(\sigma)$ with a symmetric orbital (relative to the plane bisecting the bond coordinated to the metal). These orbitals are the crossing orbitals in the symmetry-forbidden $[_\sigma 2_a + _\sigma 2_a]$ process; Ψ_2, antibonding and empty in 58, transforms to a

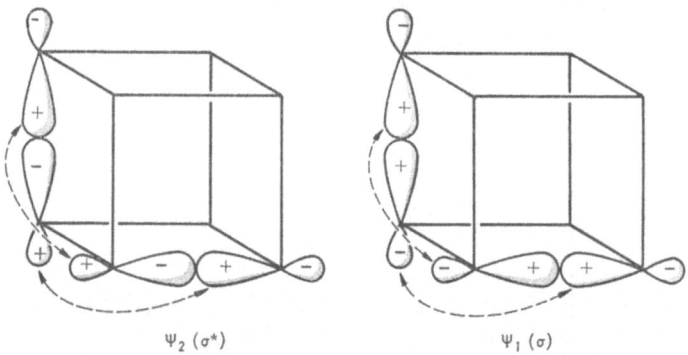

$\Psi_2(\sigma^*)$ $\qquad\qquad\qquad$ $\Psi_1(\sigma)$

bonding orbital in 59. Ψ_1 is its opposite counterpart. Catalysis may proceed through an intermediate like 17 or from simple coordination.

The nature of silver's role in the catalysis of symmetry-forbidden reactions, however, is puzzling. Silver salts are known to catalyze symmetry-forbidden electrocyclic transformations with remarkable facility [43] when rhodium(I) complexes exhibit only marginal activity [44]. Rhodium(I), of course, readily inserts into the σ bond of cubane giving the kind of intermediate required for an allowed $[_\sigma 2_s + _\sigma 2_a + _\sigma 2_a]$ transformation to the polycyclic counterpart of 59 [16], but the preferred path of transformation is to the diene *20*. In the apparent $[_\sigma 2_a + _\sigma 2_a]$ catalysis of cubyl systems, silver ions appear the more reactive [16,42a,b]. Silver salts also catalyze the valence isomerization of quadricyclene to norbornadiene, where they do not isomerize the *exo*-cyclopropyl norbornene *12* [45].

Significantly, liquid cyclopropane does not form a complex with solid silver nitrate at room temperature [46]. Further, silver salts exhibit no bonding affinity for the cyclopropane ring in *12* (*i.e.*, the equilibrium constant for complexation with norbornene was 1.6 and 0.21 for *12*; steric factors were ruled out by other means [45]. A lack of coordinate bonding affinity for cyclopropane might suggest somewhat limited back-bonding abilities by the silver ion since part of the energy gained through coordination would result from released ring strain through back-bonding. Indeed, some spectroscopic studies of the silver-olefin bond have suggested that donor bonding is more important than back-bonding [47], and direct e.s.r. evidence indicates the importance of donor overlap between the bonding alkene π orbital with a vacant s orbital on the Ag(I) ion in a series of perchlorate-cycloalkene complexes [48].

The silver ion, then, does not exhibit the same degree of back-bonding that the more familiar transition elements do. Since back-bonding is an essential factor in the forbidden-to-allowed process and, in particular, in direct oxidative addition, silver's function in this chemistry could differ. It may be that the silver ion (and other similar metallic species) stands apart from the other transition elements (W, Mo, Cr, Fe, Co, Ni, Rh, etc.) in its mode of catalysis. In the valence isomerization of quadricyclene, some oxidation occurs as evidenced by the deposition of metallic silver [45]. Certainly, irreversible redox cannot be a feature of the actual catalytic path, since silver's role is definitely catalytic and the isomerization itself precludes it (*i.e.*, the oxidation state of the system remains fixed). Some electron transfer, however, clearly proceeds and may be a critical feature of the catalysis. One could speculate on the possibility of intermediate ion radicals generated through electron transfer from a reactant to Ag(I) followed by electron recapture by the rearranged species in the catalytic system.

Silver's role in the valence isomerization of bicyclobutane to butadiene, however, is most striking and deserves special attention. Masamune *et al.*, studying the silver perchlorate-catalyzed valence isomerization of *exo,exo*- and *endo,exo*-2,4-dimethylbicyclobutane, reported largely stereospecific conversion to *trans,trans*- and *cis,trans*-2,4-hexadiene, respectively [49]. At 5 °C, the isomerization of the *exo,exo*-bicyclobutane was 77% stereospecific, while the *endo,exo* isomer was 99% specific. Significantly, both isomers rearranged along the symmetry-forbidden $[_\sigma 2_a + _\sigma 2_a]$ path in preference to alternative, symmetry-allowed $[_\sigma 2_s + _\sigma 2_a]$ paths. These results appear to be the first to demonstrate that a metal ion can inject stereospecificity into systems where stereochemical alternatives exist and, in fact, reverse the stereochemical course of the thermal process. The suggestion of the authors — namely that the specificity of the reaction is to a large measure governed, at some stage, by the orbital symmetry of the whole system undergoing skeletal change — is difficult to avoid.

Paquette et al., examining the same silver-catalyzed reaction as Masamune, reported similar stereospecificity [50]. A kinetic study of this reaction further established the reversible intervention of a silver-bicyclobutane complex. The kinetic data did not bear on the nature of the complex. The reported equilibrium constant ($K = 0.19 \pm 0.09$), however, revealed that the extent of complex formation was substantial.

Although cations or cation radicals are suggested in some silver-catalyzed systems [45,51], they do not appear to intervene here. The stereo-specificity of the bicyclobutane isomerization is inconsistent with a *free* ionic species which would be expected to transform along the thermal $[_\sigma 2_s + _\sigma 2_a]$ path. A forbidden-to-allowed role has been suggested for silver ion in its catalysis of highly strained cyclobutene ring systems [43]. However, other distinctly different modes of catalysis have been offered for silver [51,52]. A number of factors would seem to mediate against a forbidden-to-allowed role for silver in the catalysis of forbidden reactions. Chief among them are its poor coordinate bonding properties, discussed earlier, and its electronic configuration [12]. Whatever the case, silver is most intriguing in its catalytic behavior exhibiting catalytic properties placing it in a class apart from most other transition elements.

VI. Concluding Remarks

We have attempted in this chapter and in previous publications [2,3,20,29] to describe some broad guidelines associated with the metal-catalysis of symmetry-forbidden reactions. These treatments should help to better understand this very interesting, but complex, body of chemistry and to direct the attention of experimental chemists to more catalytically active systems. Although this chapter contains some new information about this, it is by no means comprehensive. The application of symmetry principles to specific reactions (particularly [2 + 2] processes) should not be considered limiting. The ideas expressed should find generality in applications in the catalysis of all symmetry-forbidden processes. The orbital symmetry factors discussed can undoubtedly be expressed in other ways. The significance of orbital symmetry concepts is only just beginning to have an impact in inorganic chemistry [53]. Its importance in the understanding of these systems should continue to grow.

The catalysis of symmetry-forbidden reactions is currently of broad interest and a number of workers have expressed different points of view on the metal's role. Katz, for example, seems to take the position that these catalytic processes are very likely stepwise in nature, and, therefore, forbidden-to-allowed arguments do not apply [15,40]. This criticism is perhaps more broadly directed toward the concerted, "π complex multicenter reaction" mechanism first postulated by Schrauzer to explain metal-assisted cycloaddition chemistry [54]. A preference for nonconcerted mechanisms in this chemistry is also shared by Cassar, Eaton, and Halpern [16].

Van der Lugt also views the catalysis of symmetry-restricted reaction differently [12]. The concerted character of the ligand transformation is not questioned, but the relevance of orbital symmetry conservation in the metal's role is. Focusing on a formally symmetry-forbidden ligand transformation, van der Lugt argues that reaction proceeds because of a lowered energy barrier due to configuration interaction. One example was a symmetry-forbidden opening of a cyclobutane ring to a bisolefin system. There existed, however, an alternative direction for ring opening which was symmetry-allowed [55]. We have discussed here the alternative routes to ring-opening available to metal-coordinated cyclobutane rings. Certainly, the energy barrier to opening along the forbidden path would be lower (from configuration interaction) than that for opening in the absence of the metal. But whether it would be sufficiently lowered to introduce reaction along that path in competition to the symmetry-allowed path is an open question. For nonsymmetrical cyclobutane ring systems (*e.g.*, quadricyclene and prismane), the symmetry-allowed path would unquestionably dominate. Here, as we have noted, coordinate bonding to the metal creates in the ligand a strong propensity to transform along the forbidden-to-allowed path, essentially locking high energy ligands to that mode of transformation. Most other symmetry-restricted [2 + 2] processes involve ligand systems with clear centers of bidentate character where alternative (formally forbidden) modes of transformation are obscure. We do not question the importance of configuration interaction where symmetry barriers to reaction exist. The central point of difference, however, involves the relevance of orbital symmetry factors in explaining the lowered energy barriers to forbidden reactions and the necessary connection between orbital symmetry conservation and the catalysis of these reactions.

The forbidden-to-allowed process, however, is an outgrowth of ligand-to-metal coordinate bonding. The components of the coordinate bond open the forbidden-to-allowed path of reaction to ligand systems. This can be expressed in another way. Two olefins, for example, coordinated to a transition metal capable of strong back-bonding, experience a molecular distortion in the direction of the corresponding cyclobutane. The contribution of the excited state configuration of the bisolefin system lends cyclobutane character to the coordinated system, which may be expressed by a contributing structure *(61)* [29].

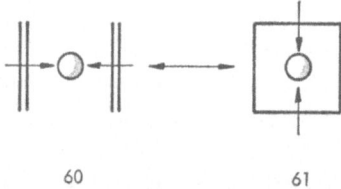

60   61

If the metal system in [*60* ↔ *61*] possesses a restrictive ligand field, the ligand will retain greater olefin character (*i.e.*, *60* makes the major contribution). In the absence of restrictive ligand fields, the bisolefin system can assume any bonding configuration between the extremes represented by the contributing structures, adopting a structural mix dictated by the energy factors of the bonding system. More important, it is in the nature of the coordinate bond to create a propensity for ligand transformation along the forbidden-to-allowed path with preservation of full bidentate coordinate bonding. Quadricyclene, for example, directing its preferred centers of coordinate bonding to bidentate sites of coordination on a metal like *61*, will experience molecular distortion in the direction of norbornadiene due to the mixing-in of contributing structure *60*. Complete molecular relaxation can proceed in that direction yielding a norbornadiene ligand fully coordinated to the metal center. Clearly, coordinate bonding and symmetry conservation are tightly related. Orbital symmetry conservation would only seem unimportant to the metal's role if coordinate bonding itself is not significant to this catalysis. If full coordinate bonding is important in the chemistry of these systems, then orbital symmetry conservation becomes equally so; the two factors are inseparably related.

Woodward and Hoffmann have recently invited speculation on the possible existence of *trans*-bicyclo[2,2,0]hexadiene *(62)* [1]. The *cis*-isomer *63* is a relatively stable compound owing its existence to the fact that the only geometrically simple paths of rearrangement to its preferred valence isomer benzene are symmetry-forbidden. The question about the possible existence

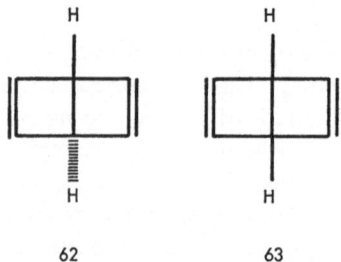

62   63

of *62*, of course, goes to the heart of orbital symmetry conservation, sharply contrasting those molecular systems locked in bonding configurations by symmetry restrictions to those with molecular freedom. *62* is free to transform to benzene through a symmetry-allowed, conrotatory electrocyclic process. It could, therefore, not exist at all as a distinct molecular species, but could rather represent a point along the energy surface of benzene.

The capacity of a metal system to remove orbital symmetry restrictions invites similar speculation on the survival capabilities of certain metal-coordinated species. The existence, for example, of quadricyclene coordinated to certain transition metals [29] in a bidentate configuration rests upon the ligand field restrictions described by the nonreacting ligands. When these restrictions are removed, the existence of quadricyclene becomes as tenuous as that of *trans*-bicyclo[2,2,0]hexadiene.

References

[1] Seebach, D.: Topics Curr. Chem. *11*, 177 (1969). — Hoffmann, R., Woodward, R. B.: Science *167*, 825 (1970). — Woodward, R. B., Hoffmann, R.: Angew. Chem. Intern. Ed. Engl. *8*, 781 (1969). — Fukui, K.: Topics Curr. Chem. *15*, 1 (1970).

[2] For a general discussion and review of some of this chemistry, see Mango, F. D.: Advan. Catalysis *20*, 291 (1969).

[3] Mango, F. D., Schachtschneider, J. H.: J. Am. Chem. Soc. *89*, 2484 (1967).

[4] Horiuti, I., Palanyi, M.: Trans. Faraday Soc. *30*, 1164 (1934). — Gault, F. G., Rooney, J. J., Kemball, C.: J. Catalysis *1*, 255 (1962).

[5] Pettit, R., Sugahara, H., Wristers, J., Merk, W.: Discussions Faraday Soc. *47*, 71, (1969).

[6] Cowherd, F. G., von Rosenberg, J. L.: J. Am. Chem. Soc. *91*, 2157 (1969).

[7] Dewar, M. J. S.: Bull. Soc. Chim. France *C 71* (1951). — Chatt, J., Duncanson, L. H.: J. Chem. Soc. *1953*, 2939.

[8] Belluco, U., Crociani, B., Pietropaolo, R., Uguagliati: Inorg. Chem. Acta *3*, 19 (1969). — Collman, J. P., Kang, J. W.: J. Am. Chem. Soc. *89*, 844 (1967). — Chatt, J., Shaw, B. L., Williams, A. A.: J. Chem. Soc. *1962*, 3269.

[9] a) Banks, R. L., Bailey, G. C.: Ind. Eng. Chem., Prod. Res. Develop. *3*, 170 (1964); b) Calderon, M., Ofstead, E. A., Ward, J. P., Scott, K. W.: J. Am. Chem. Soc. *90*, 4133 (1968).

[10] Hogeveen, J., Volger, H. C.: J. Am. Chem. Soc. *89*, 2486 (1967).

[11] Ferguson, L. H.: J. Chem. Educ. *47*, 46 (1970).

[12] van der Lugt, W. Th. A. M.: Tetrahedron Letters *1970*, 2281.

[13] Volger, H. C., Hogeveen, H., Gaasbeek, M. M. P.: J. Am. Chem. Soc. *91*, 218 (1969).— Katz, T. J., Cerefice, S. H.: Tetrahedron Letters *1969*, 2509, 2561.

[14] Bailey, N. H., Gillard, R. D., Keeton, M., Mason, R., Rusell, D. R.: Chem. Commun. *1966*, 396. — Irwin, W. J., McQuillin, F. J.: Tetrahedron Letters *1968*, 1937.

[15] Katz, T. J., Cerefice, S.: J. Am. Chem. Soc. *91*, 6519 (1969).

[16] Cassar, L., Eaton, P. E., Halpern, J.: J. Am. Chem. Soc. *92*, 3515 (1970).

[17] Dauben, W. G., Kielbania, A. J., Jr.: J. Am. Chem. Soc. *93*, 7345 (1971).

[18] Collman, J. P.: Accounts Chem. Res. *1*, 136 (1968). — Chock, P. B., Halpern, J.: J. Am. Chem. Soc. *88*, 3511 (1966). — Halpern, J.: Advan. Chem. Ser. *70*, 1 (1968).

[19] a) Longuet-Higgins, H. C., Orgel, L. E.: J. Chem. Soc. *1959*, 1956;
b) see also Hübel, W.: In: Organic synthesis via metal carbonyls (ed. I. Wender and P. Pino), Vol. I. New York: Interscience Publishers 1968.

[20] Mango, F. D., Schachtschneider, J. H.: J. Am. Chem. Soc. *91*, 1030 (1969).

[21] Hübel, W., Brayl, E. H., J. Inorg. Nucl. Chem. *10*, 250 (1959). — Nakamura, A. Hagihara, N.: Bull. Chem. Soc. Japan *34*, 452 (1961).

[22] Krüerke, U., Hübel, W.: Chem. Ber. *94*, 2829 (1961).

[23] Reviews: Bowden, F. L., Lever, A. B. P.: Organometal. Chem. Rev. A *3*, 227 (1968). — Maitlis, P. M.: Advan. Organometal. Chem. *4*, 95 (1966).

[24] Zeiss, H., in: Organometallic chemistry (ed. H. Zeiss), p. 380; American Chemical Society Monograph No. 147. New York: Reinhold Publishing Corp. 1960. — Arnett, E. M., Bollinger, J. M.: J. Am. Chem. Soc. *86*, 4729 (1964).

[25] Whitesides, G. M., Ehmann, W. J.: J. Am. Chem. Soc. *91*, 3800 (1969).

[26] Penella, F., Banks, R. L., Bailey, G. C.: Chem. Commun. *23*, 1548 (1968).

[27] Eisch, J. J., Amtmann, R., Foxton, M. W.: J. Organometal. Chem. *16*, 55 (1969). — Schäfer, W.: Angew. Chem. *78*, 716 (1966).

[28] Muetterties, E. L.: Accounts Chem. Res. *3*, 266 (1970).

[29] a) Mango, F. D.: Chem. Technol. *1*, 758 (1971);
b) Mango, F. D.: Tetrahedron Letters *1973*, 1509.

[30] Volger, H. C., Gaasbeek, M. M. P., Hogeveen, H., Vrieze, K.: Inorg. Chim. Acta *3*, 145 (1969).

[31] Pettit, R.: J. Am. Chem. Soc. *81*, 1266 (1959). — Lemal, D. M., Shim, K. S.: Tetrahedron Letters *1961*, 368. — Trecker, D. J., Foote, R. S., Henry, J. P., McKeon, J. E.: J. Am. Chem. Soc. *88*, 3021 (1966).

[32] Jennings, W., Hill, B.: J. Am. Chem. Soc. *92*, 3199 (1970).

[33] Strohmeier, W.: Angew. Chem. Intern. Ed. Engl. *4*, 730 (1964).

[34] Reppe, W., Schlichting, O., Klager, K., Towpel, T.: Justus Liebigs Ann. Them. *560*, 1 (1948).

[35] Schrauzer, G. N.: Angew. Chem. Intern. Ed. Engl. *3*, 185 (1964).

[36] For a review of this subject see Schrauzer, G. N.: Advances in organometallic chemistry (ed. F. G. A. Stone and R. West), Vol. 2. New York: Academic Press 1964.

[37] Hoogzand, C., Hübel, W.: Organic synthesis via metal carbonyls (ed. I. Wender and P. Pino), Vol. 1, p. 343. New York: Interscience Publishers.

[38] Schrauzer, G. N., Bastian, B. N., Fosselins, G. A.: J. Am. Chem. Soc. *86*, 4890 (1966).

[39] Schrauzer, G. N., Ho, R. K. Y., Schlesinger, G.: Tetrahedron Letters *1970*, 543.

[40] Katz, T. J., Acton, N.: Tetrahedron Letters *1967*, 2601.

[41] Boer, E. P., Tsai, J. H., Flynn, J. J., Jr.: J. Am. Chem. Soc. *92*, 6092 (1970).

[42] a) Dauben, W. G., Buzzolini, M. G., Schallhorn, C. H., Whalen, D. L.: Tetrahedron Letters *1970*, 787;
b) Paquette, L. A., Stowell, J. C.: J. Am. Chem. Soc. *92*, 2584 (1970).

[43] Merk, W., Pettit, R.: J. Am. Chem. Soc. *89*, 4788 (1967).

[44] Volger, H. C., Hogeveen, H.: Rec. Trav. Chim. *86*, 830 (1967).

[45] Menon, B. C., Pincock, R. E.: Can. J. Chem. *47*, 3327 (1969).

[46] Tipper, C. F. H.: J. Chem. Soc. *1955*, 2045.

[47] Powell, D. B., Sheppard, N.: J. Chem. Soc. *1960*, 2519. — Schug, J., Martin, R. J.: J. Chem. Phys. *66*, 1554 (1962). — Quinn, H. W., McIntyre, J. S., Peterson, D. J.: Can. J. Chem. *43*, 2896 (1965).

[48] Gee, D. R., Wan, J. K. S.: Chem. Commun. *1970*, 641.

[49] Sakai, M., Yamaguchi, H., Westberg, H. H., Masamune, S.: J. Am. Chem. Soc. *93*, 1043 (1971).

[50] Paquette, L. A., Wilson, S. E., Henzel, R. P.: J. Am. Chem. Soc. *93*, 1288 (1971).

51) Kaiser, K. L., Childs, R. F., Maitlis, P. M.: J. Am. Chem. Soc. *93*, 1270 (1971).
52) Sakai, M., Masamune, S.: J. Am. Chem. Soc. *93*, 4610 (1971). — Paquette, L. A.: Accounts Chem. Res. *4*, 280 (1971). — Gassman, P. G., Atkins, T. J., Williams, F. J.: J. Am. Chem. Soc. *93*, 1812 (1971).
53) Eaton, D. R.: J. Am. Chem. Soc. *90*, 4272 (1968). — Pearson, R. G.: J. Am. Chem. Soc. *91*, 4947 (1969). — Bader, R. F. W.: Can. J. Chem. *40*, 1164 (1962). — Whitesides, T. H.: J. Am. Chem. Soc. *91*, 2395 (1969).
54) Schrauzer, G. N., Glockner, P.: Chem. Ber. *97*, 2451 (1964). — Schrauzer, G. N.: Advan. Catalysis *18*, 373 (1968).
55) Mango, F. D.: Tetrahedron Letters *1971*, 505. — Mango, F. D., Schachtschneider, J. H.: J. Am. Chem. Soc. *93*, 1123 (1971).

Received August 16, 1972

The Relationship between Mass Spectrometric, Thermolytic and Photolytic Reactivity

Prof. Ralph C. Dougherty

Department of Chemistry, Florida State University, Tallahassee, Florida, USA

Contents

I. Introduction

The analogy between the fragmentation reactions of molecule ions and thermochemical or photochemical decomposition of the same molecules has been a subject of interest since the beginning of organic mass spectrometry.[1,2] The relative ease of obtaining a mass spectrum — a few minutes work — as compared to the effort associated with the study of a given pyrolysis or photolysis reaction has provided a direct impetus for analysis of the relationships between molecular reactivity in the three systems. The literature contains numerous reports of thermochemical and photochemical reactions that were discovered because the mass spectrum of the starting material suggested that it might selectively decompose in a useful way. There are also a number of reports that illustrate apparent failures of the relationship between mass spectral reactivity and either thermochemistry or photochemistry.

In this chapter we will discuss the basis for the relationship between reactivity in a mass spectrometer and reactivity under thermal or photochemical activation. We will present an empirical guide which may be useful in predicting relationships between the three types of unimolecular reactivity. In the light of this analysis, we will review the cases where the reactivity relationships appear to have broken down; and finally, we will review the successful examples of the relationship between mass spectrometry and thermochemistry, and mass spectrometry and photochemistry.

Meyerson and Fields have reviewed the thermochemistry of aromatic systems and its relationship to corresponding mass spectra with particular emphasis on their extensive studies of thermally generated arynes.[3-5] Maccoll has reviewed the literature in the general area prior to 1966.[6] Partial reviews of unimolecular reactivity relationships have also appeared within discussions of mass spectral reactivity in general.[7,8]

It is our objective in this chapter to contribute to the development of the empirical relationships between mass spectral and other forms of unimolecular reactivity. We hope to develop an understanding of this area that is more subtle than a simple 'see it here — see it there' relationship, and yet more tractable and generally useful than a rigorous quantum mechanical or quasi-equilibrium theoretical discussion.

II. The Basis for the Relationships between Mass Spectrometric, Thermochemical and Photochemical Reactivity

In our discussion of the foundations of the relationship between mass spectral, thermal and photochemical reactivity we will use a perturbational approach.[9-12) There are three kinds of perturbations that will influence the reactivity in the three systems, namely

(1) electronic state,
(2) charge, and
(3) energy content and distribution.

As long as all of the reactions occur in the gas phase, we need not consider medium effects. For thermolysis or photolysis reactions that occur in condensed phases the medium may substantially alter the reactivity and must be considered in discussion of analogies. We will discuss each of the perturbations above in the following paragraphs.

1. Effect of Electronic States

The ground state of a molecule ion can be thought of as a hybrid of the ground state and first excited state of the molecule. Fig. 1 illustrates the occupancy of the highest occupied and lowest vacant molecular orbitals

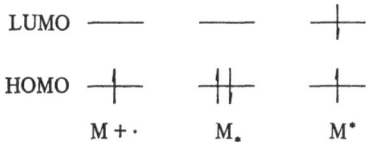

Fig. 1. Orbital occupancy in molecule ions ground states and first excited states

Ionization of a molecule generally does not substantially perturb the molecular vibrational energy levels [13), and as Brown and his coworkers have pointed out, the parallelism of bond energies and vibrational modes in molecules and their corresponding ions should result in a parallelism between mass spectral and thermal fragmentation reactions. [14) The details of the relationship will, however, be strongly influenced by the nature of the reactive electronic states that are involved. The strong dependence of chemical reactivity on the nature of the reactive electronic states is well known.[9,10) The analysis above suggests that mass spectral reactivity should be a hybrid between thermal and photochemical processes and caution should be exercised in anticipating parallels between mass spectral and thermal or photochemical reaction courses.

The problem becomes even more complicated if one is willing to consider the possibility of reactions from excited electronic states in the mass spectrometer. Mass spectral reactions that occur from electronically excited states of the molecule ion are known [7,8) and to the extent that electronic states control reactivity these states must be considered. Photochemical reactions differ from thermal processes in two important respects. First, electronic excitation usually results in the selective activation of a limited number of degrees of freedom in the molecule so that photochemical reactions show different selectivity than their thermochemical counterparts. Second, photochemical reactions tend to proceed through "Bell-Evans-Polanyi" unstabilized transition states. [9,10,12] This distinction is significant only in the case of pericyclic reactions. [9-12,15] We proposed that metastable ions be used to distinguish reactions that occurred in the electronic ground state from those that probably had excited state precursors. [11] If a mass spectral pericyclic reaction gives a significantly intense metastable ion ($\sim 10^{-4} \times$ daughter ion intensity) in a sectror mass spectrometer the reaction probably occurs from the ion's ground state and should be analogous to a thermal pericyclic reaction. If no metastable is observed, the reaction may have occurred from an excited state of the ion and should be analogous to a photochemical pericyclic reaction.

This analysis is very satisfactory for the retro-Diels-Alder reactions, e.g. the decomposition of vinyl cyclohexene (1) [16] which are ground state

$$\text{(1)}$$

pericyclic processes. Unfortunately many if not most of the pericyclic reactions that occur in molecule ions are rearrangement reactions and the time based metastable ion analysis cannot be unequivocally applied to these processes. For example, the hexahelicene rearrangement (2) [11] is directly analogous to the thermal pericyclic reaction of the neutral. The

$$\text{+ C}_2\text{H}_4 \qquad \text{(2)}$$

cyclization of stilbene (3) and diphenylmethyl cations (4) follows an excited state path [17], nonetheless there are prominent metastable ions in the spectra for loss of hydrogen or hydrogen atoms. [17] Similarly the mass spectra of the muconic and cyclobutene dicarboxylic acids are related in such a way that the ions must equilibrate through a "photochemical" pericyclic pathway prior to fragmentation or metastable ion decomposition [18] (5), (6).

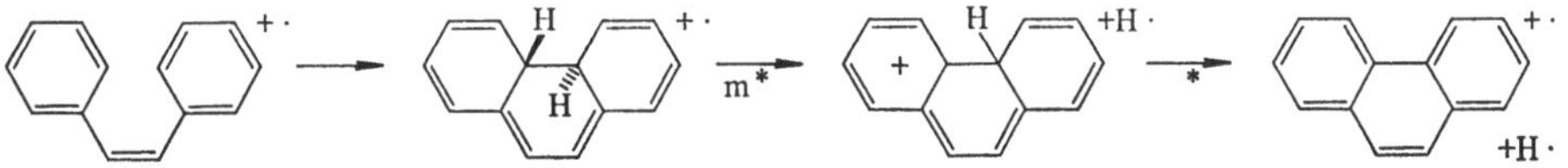

(3)

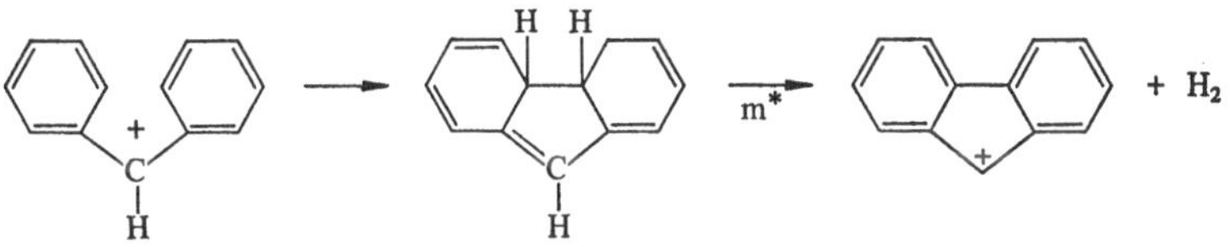

(4)

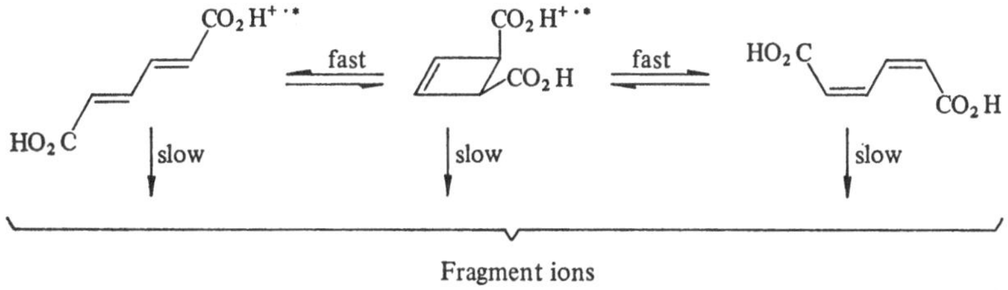

Fragment ions

(5)

Fragment ions

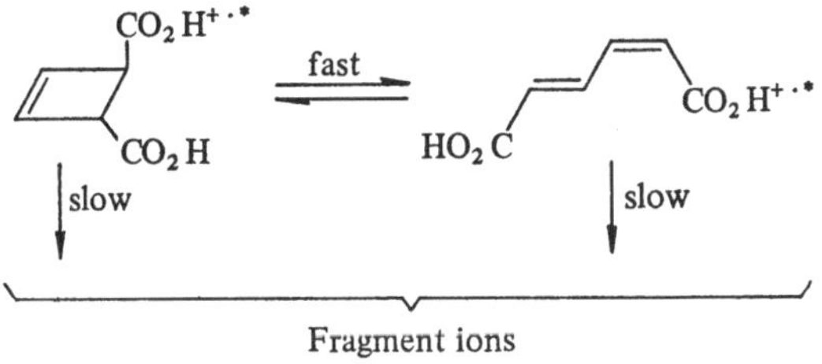

(6)

The well known scrambling of carbons and hydrogens that occurs in many benzenoid systems prior to fragmentation is likewise an example of an electron impact analog of a photochemical reaction. In all of these cases establishment of a correlation depends on first determining that a rearrangement has occurred and secondly deducing the stereochemistry of the reaction.

In the case of nonpericyclic reactions the determination of whether a given reaction in a mass spectrometer should have a thermal or photochemical analog will depend on whether the particular degrees of freedom associated with the reaction can be conveniently activated by light, and the anticipated effect of the ionic charge on reactivity (see below).

The energy gap between the ground state and the first excited state of a molecule ion (~ 1 eV) is characteristically smaller than the corresponding gap for the neutral ($\sim$3—4 eV). [13] Thus if the appearance potential of a fragment ion is of the order of 2—3 eV above that of the molecule ion, the reaction can be safely considered as occurring from either the ground or excited state of the molecule as both states will be involved at those energies. This is fundamentally why the products of electron impact, high energy thermal, photochemical and radiochemical reactions are usually quite similar.

2. Effect of Ionic Change

One of the largest differences between fragmentation in a mass spectrometer and fragmentation of neutrals is the charge associated with the fragmenting ions. From the point of view of perturbation theory [9,10] it is convenient to analyze charge effects on reactivity for hydrocarbons and for systems with heteroatoms separately.

Removal of one electron should make no difference to the relative stabilities of polyene molecule ions or even electron polyene fragments as compared to their neutral counterparts, e.g. butadiene and the allyl radical should have the same relative stabilities as the butadiene molecule ion, and the allyl cation. Removal of one electron will, however, alter the stabilities, and thus the reactivities of cyclic polyenes. The molecule ions of aromatic hydrocarbons will be substantially less aromatic then their neutral counterparts.[11] Correspondingly the molecule ions of antiaromatic hydrocarbons will not be as antiaromatic as their neutral analogs, e.g. cyclobutadiene$^{+ \cdot}$ should be relatively more stable than cyclobutadiene. The largest charge effects in hydrocarbons will be observed in nonalternant[a] monocyclic hydrocarbons. The cyclopropenium ion *1* and the tropillium ion *2* are both strongly aromatic as compared to their neutral analogs. Consequently $C_3H_3^\oplus$ is a very common ion in the mass spectra of hydrocarbons while cyclopropene is not a common product of hydrocarbon pyrolysis or photo-

---

[a] *Alternant* hydrocarbons π-electron systems can be starred in such a way that no two starred or unstarred atoms are connected to members of a like set. *Odd* or *even* refer to the number of atoms in the π-system. Nonalternant hydrocarbons cannot be so starred, e.g.

Even alternant   Odd alternant   Nonalternant

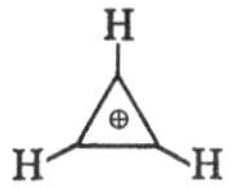

1 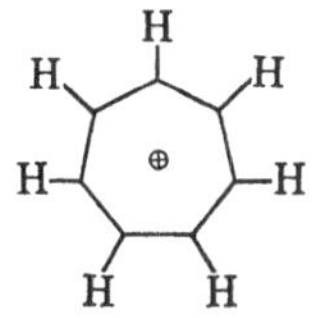   *2*

lysis. Similarly the $C_7H_7^+$ ion is very frequently a tropillium ion [18] in the mass spectra of aromatic hydrocarbons while pyrolysis experiments typically yield products that are more directly related to benzyl radicals.[19]

Introduction of heteroatoms complicates the analysis of charge effects on fragmentation reactions. In addition to the consideration of aromaticity effects discussed above, heteroatoms have profound effects on the stabilities of ions which contain them. There are essentially three cases that merit consideration: (1) odd electron ions; (2) odd alternant* (see footnote p. 98) even electron ions; and (3) even alternant* (see footnote on p. 98) even electron ions.

a) Odd Electron Ions

There should be relatively little effect on the relative stability of odd electron ions that contain heteroatoms, although the relative stability of their even electron fragments may be considerably altered. The absence of a hetero-atom effect on odd electron → odd electron fragmentation reactions is clearly illustrated by the McLafferty rearrangement. In the mass spectro-

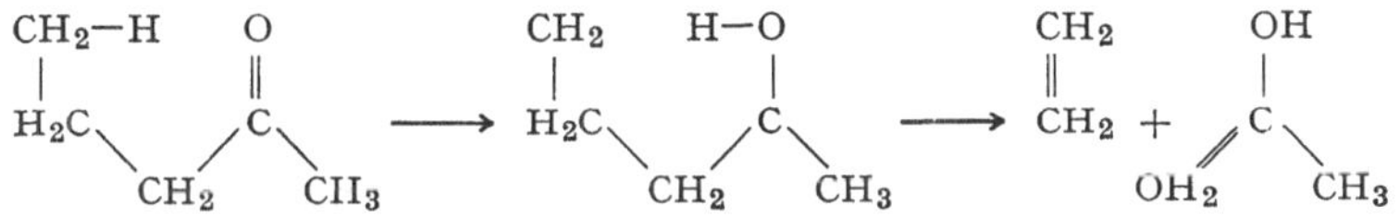

$$\tag{7}$$

meter reaction (7) is known as the McLafferty rearrangement. In photochemistry the same reaction is called the Norrish *type II* photochemical cleavage, and in thermochemistry the reaction is called a γ-hydrogen rearrangement. The reactivity is similar in all three systems which is consistent with a very small charge effect on the reactivity in this process. The Norrish type II eliminations from 4-phenyl-4-methyl-2-pentanone, *3*, are substantially slower (0.4:1) than corresponding reactions of 2-pentanone, *4*.[20]

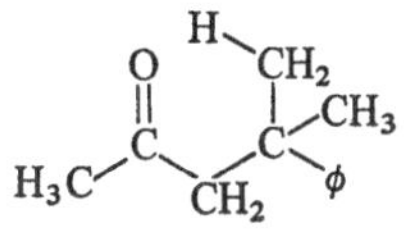

3 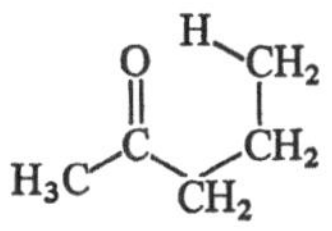   *4*

The McLafferty rearrangement of *3* gives an ion of 119% relative intensity (compared to the molecule ion) which has been interpreted as indicating a failure of the analogy between photochemistry and mass spectrometry [20]; the McLafferty rearrangement ion from 2-pentanone, *4*, is 64% of the molecule ion intensity.[21] The difference between the results of these two reactivity comparisons may be due to a charge effect on the reactivity of the heteroatomic systems. It seems to us, however, that the effect is a quantitative one and not a qualitative one; the basis of the effect is most likely the difference in energy deposition between photolysis and electron impact. When we consider all of the factors that can influence correlations between photochemistry and mass spectrometry, quantitative correlations are not expected. Indeed the qualitative correlation between the reactivity of odd electron heteroatomic systems is rewarding enough; at times even this correlation appears to break down (see below).

b) Odd Alternant Even Electron Ions

In order for a heteroatom to have a first order effect on the stability of odd alternant ions the atom must be located at an active[b] site in the ion. If the electronegativity of the heteroatom is less than that of carbon the cation, will be stabilized. The reverse is true if the heteroatom is more electronegative than carbon.

The mass spectrum of 4-hydroxy-3-methoxy-4,3-borazaroisoquinoline *5*, shows a relatively very abundant M-1 ion (60%). The presumed cleavage ion, *6*, is stabilized relative to related molecules because boron is an active atom in the cation. We would *not* expect the thermolysis of *5* to produce

5   *6*

products related to M-1 to nearly the extent that they occur in the mass spectrum of *5*.

The mass spectrum [22] and thermolysis [23] of cyclohexanone illustrates the effect of heteroatoms on odd alternant hydrocarbon stability when the heteroatom is more electronegative than carbon. In the mass spectrum of cyclohexanone virtually all of the fragment ions appear to be daughters of the initial α-cleavage ion *7*.[22]

---

[b] *Active* sites are the starred atoms in an odd alternant hydrocarbon when the starred set of atoms is more numerous than the unstarred set.

7   8   9

In the thermolysis of cyclohexanone virtually all of the products arise from the β-cleaved structure 8.[23] Two effects are important here. In the ion the α-cleavage structure is specifically stabilized (see below) and the β-cleavage structure 9, would be specifically destabilized by the heteroatom which is at an active site in the odd alternant π system associated with the carbonyl. The primary photolysis products of cyclohexanone [24] are much more closely correlated to its mass spectrum than the thermolysis products are. This is because n→π* excitation can be relaxed by an α-cleavage (Norrish type I) analogous to 7. The analogy for the photochemical reactions is, however, far from perfect because of the stabilizing effect of the oxygen atom on the even alternant even electron ions, e.g. 10. The photolysis of

10

cyclohexanone results primarily in decarbonylation products [24] whereas most of the ions in the mass spectrum contain the oxygen atom.[22]

c) Even Alternant Even Electron Ions

The example cited above is one case where charge effects in even alternant even electron heteroatomic ions have muddled the correlation between fragmentation in a mass spectrometer and other types of reactivity. The reason for the lack of correlation stems from the fact that ions in this group generally have one more bond, or at least a half bond more than their neutral counterparts. The stabilization conferred by the extra bond invariably outweighs the destabilizing effect of placing a positive charge on an electronegative heteroatom. Structure 10 is one example of this type of ion; others are contained in the ions that result from α-cleavage, e.g. 11 and 12 and internal displacement reactions, e.g. 13 and 14. Since these ions are specifically stabilized as compared to their neutral odd electron analogs we would not universally expect products related to these structures to be important in the thermochemistry or photochemistry of the molecules

$$H_2C{=}O^+{-}H \qquad\qquad H_2C{=}\overset{+}{N}{\raise2pt\hbox{$\scriptstyle\diagup$}}{\raise-2pt\hbox{$\scriptstyle\diagdown$}}\genfrac{}{}{0pt}{}{CH_3}{CH_3}$$

11   *12*

13   *14*

concerned. Structures that may be represented as protonated molecules, *e.g. 11* and *13*, will probably have analogs in neutral unimolecular reactions while the even electrons *10, 12* and *14* and related structures will generally be unique to mass spectrometry. It is conceivable that *14* could be related to a thermochemical product like *15*; however, this seems unlikely on chemical grounds.

15

3. Effect of Energy Content and Distribution

The energy content and the distribution of energy in the reacting system is dramatically different when one compares mass spectrometry, thermo-chemistry and photochemistry. The discussion below shows that in spite of the large differences in energy distribution between the three systems the relative reactivity should be only slightly effected by changing excitation methods.

Although there is still a considerable controversy concerning the energy deposition function for 70 eV electron impact on molecules, most investigators would agree that the internal energy of molecule ions formed at 70 eV may extend up to 10 eV. That is nearly twice the energy that is available by conventional photochemistry or thermochemistry. The electron impact energy deposition function depends on the transition probability for formation of ions with specific interval energies; in contrast to a Maxwellian thermal distribution, this function will generally have considerable structure and be sharply peaked at particular energies.

The energy deposition function for photochemical reactions is given by absorption spectrum of the substrate. This function is of course unique when compared to the corresponding functions for thermal or electron impact activation.

With the exception of photochemical reactions in which specific degrees of freedom are activated by photon impact, the energy distribution function should not substantially influence relative reactivity. Subject to the exceptions listed above the mass spectrum of a compound will probably contain ions representative of its photochemistry throughout its absorption spectrum, it will also contain thermochemically related ions. The reason why the energy deposition function isn't very important to relative reactivity stems from the fact that relative reactivity is primarily controlled by the "frequency factors" and activation energies for the set of reactions. Increasing the energy content of the reactant will increase the rate of all of the reactions. In the extreme, selectivity between reactions with low activation energies, and those with high activation energies, will disappear but both processes should still occur. The very high energy reactions that occur in a mass spectrometer can easily be eliminated from consideration by appearance potential measurements or studies at low voltages.

III. Qualitative Guides for Anticipating Correlations between Mass Spectrometric, Thermal and Photochemical Reactions

(i) Only the *primary products* of photolysis or thermolysis experiments can be anticipated by examining a given mass spectrum. Thus the primary reaction products should be preserved to the fullest extent possible. Generally this means that the thermolysis reactions should be carried out in the gas phase with short contact times.

(ii) Consider only the *most intense ions*, and *interpret the mass spectrum* using conventional organic chemical reasoning.[22] By considering only the upper 70—80% of the total ionization in a mass spectrum, the interpretation of the spectrum is substantially simplified and it is likely that you will have considered all of the ions in the spectrum that contain relevant information about thermochemical [19] or photochemical reactivity. The success of mechanistic interpretations of organic mass spectra is indeed the foundation on which mass spectrometric-thermochemical and photochemical correlations rest.

(iii) The *energy available* for reaction and the *relative selectivity* should be considered. Reactions in a mass spectrometer usually have lower activation energies than their neutral counterparts. If the activation energy correspondence were direct, a reaction with a 3 eV activation energy in the mass spectrometer would be observable in thermolysis at 800 °C ($t_{1/2} = 0.1$ sec, Arrhenius preexponential factor $A = 10^{16}$), and photolysis at 400 nm.

Thermolysis and photolysis reactions are both more selective than ionic fragmentation processes. This is because there is genrally more energy available to ions in a mass spectrometer and the activation energies for the

103

ionic reactions are usually lower than for corresponding reactions of neutrals. This means that the extensively decomposed or rearranged ions that are common in many mass spectra will very likely not have conventional thermal or photochemical analogs. For example, in the mass spectrum of cyclohexanone [22], deuterium labeling has shown that reaction (8) occurs. The neutral which corresponds to 16 is prominent in the photochemistry

16

(8)

17

of cyclohexanone [24], however, cyclohexanone photochemistry lacks analogs of 17. In addition to the fact that 17 represents an extensive re-arrangement-fragmentation process the ion, 17, is a specifically stabilized even electron even alternant ion.

(iv) Ions that correspond to *aromatic cations*, e.g. $C_3H_3^+$ should be relatively unimportant in the thermochemistry or photochemistry of the system under study. The same is true for many *even alternant even electron heteroatomic ions*, although generalizations in this area are subject to many exceptions. For example, ions resulting from α-cleavage of ketones have analogs in both the photochemistry and thermochemistry of ketones; whereas the ion 17 above is without an analog in the photochemistry of cyclohexanone. The even electron even alternant heteroatomic ions that result from α-cleavage will generally be more directly related to the photo-chemistry $(n \rightarrow \pi^*)$ than the thermochemistry of the molecule in question.

(v) Low activation energy reactions $(E^\dagger < 1 \text{ eV})$ that do not show prominent metastable ions $(Im^*/I \text{ daughter} \sim 10^{-4})$ are likely to be more directly related to the photochemistry of the molecule than its thermo-chemistry.[11] The converse is also true.

(vi) There are a number of thermochemical reactions that will not have analogs in the compound's mass spectrum because the required ionic structures would be specifically destabilized by the requirement that a heteroatom attain a substantial positive charge (see the discussion of cyclohexanone above.)

IV. Exceptions

The exceptions to the analogy between mass spectral fragmentation and thermal or photochemical reactions are important as they indicate the limits beyond which qualitative arguments like those above will break down. In general the reported exceptions to the correlation of ionic fragmentation and thermal or photochemical reactivity can be divided into four classes. The first class includes those exceptions that are the result of heteroatom effects on the relative stabilities of ions and neutrals. Since these effects are different for thermal and photochemical reactions many of the exceptions to mass spectrometric-thermal correlations should have photochemical analogs, cf. cyclohexanone above. The second group of exceptions arise from steric effects on the relative stabilities of neutrals and ions. The third group of exceptions may be generally regarded as nonquantitative correlations. In these cases reactivity in the mass spectrometer is only partially reflected in the thermochemistry or photochemistry. This can be the result of heteroatom effects and/or the low selectivity of reactions in a mass spectrometer. The fourth group of "exceptions" are actually parallels in reactivity in which the ions in the mass spectrometer have one hydrogen atom more or less than the anticipated neutral fragment. We will discuss each of these four cases with appropriate examples from the literature.

1. Heteroatom Effects

One of the most striking apparent breakdowns in the reactivity analogy is the ionic and thermal fragmentation of o-nitroanisole, 18, which has been reported by Lossing and his coworkers.[25] Eq. (9) illustrates the major ionic

105

fragmentation processes for *o*-nitroanisole.[25] The major thermal fragmentation paths for *18* are illustrated in Eq. (10). The thermolysis products with

$$(10)$$

molecular weights of 122, 92 and 94 Daltons, *19, 20* and *21* respectively, could have been anticipated on the basis of the mass spectrum of *18*; in the mass spectrometer the ions corresponding to *19* and *21* were protonated (ions at m/e 123 and 95 respectively). The thermolysis product with a mass of 106 Daltons, *22*, could not be anticipated on the basis of the mass spectrum. This is most likely due to heteroatom destabilization of the corresponding ions as suggested by the original authors.[25] The absence of thermolysis products that would be related to the ions at m/e 123 ($C_6H_5NO_2$) and m/e 106 (C_6H_4NO) could reflect specific heteroatom stabilization of the corresponding ionic structures. On the basis of the photochemical reactivity of other aromatic nitro compounds [24], it seems likely to us that photolysis of *18* would yield products that were related to the ions at m/e 123 and 106 in the mass spectrum of *o*-nitroanisole.

The mass spectrum and thermochemistry of salicilaldehyde [26] shows many parallels to the case above. The major ionic fragmentation reactions of salicilaldehyde, *23*, are illustrated in Eq. (11). The thermal fragmentation pattern of the same molecule is illustrated in Eq. (12).[26] The ionic fragmentation that proceeds from m/e 121, Eq. (11), directly parallels the thermal reactivity of salicilaldehyde, Eq. (12); whereas the ions at m/e 107 and

m/e 76 do not have analogs in the thermolysis of 23. This "exception" may be due to specific heteroatom stabilization of the ion at m/e 104. The reaction which leads to m/e 104 would be analogous to a $n \rightarrow \pi^*$ process in the aldehyde on this basis, we would expect the photochemistry of salicilaldehyde to show products related to $(M-H_2O)$.

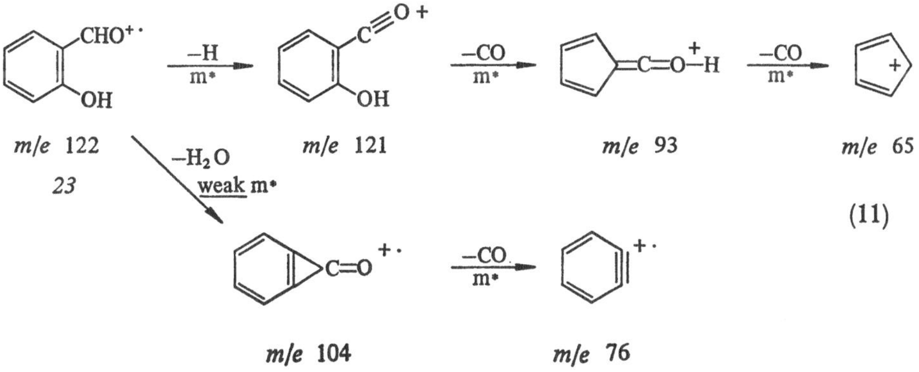

(11)

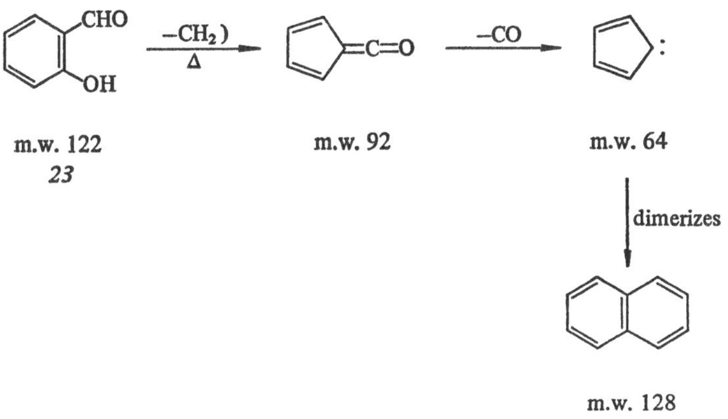

(12)

A number of cases have been reported in which the thermal fragmentation and ionic decomposition of heterocycles proceed by unique paths. Most of these divergences appear to be specific heteroatomic effects.

Cotter and Knight have reported the ionic and thermal fragmentation of 2,5-diphenyl-1,3,4-oxadiazole [27], 24. The major fragmentation pathways for the molecule ion [28] and neutral [27] are shown in Eq. (13) and (14) respectively. The benzoyl cation (m/e 105) is certain to be relatively more stable than its neutral analog; the reverse is true of the benzonitrile molecule. It seems likely that products which result from loss of H_2, including benzoyl radicals will be obtained from 24 on photolysis.

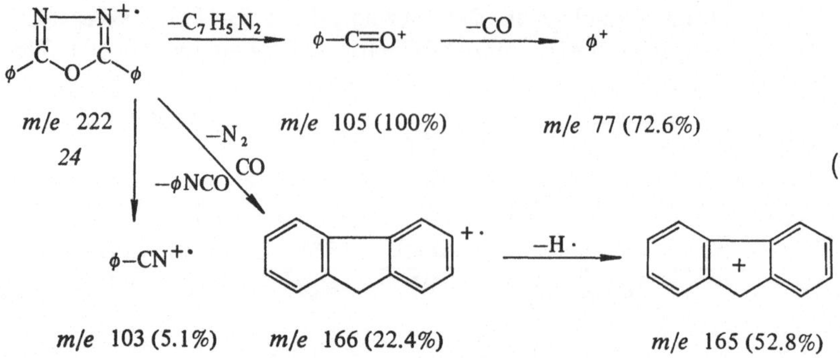
N—N+·
−C7H5N2
ϕ—C≡O+
−CO
ϕ+
m/e 222
24
−N2
CO
−ϕNCO
m/e 105 (100%)
m/e 77 (72.6%)
ϕ–CN+·
−H·
m/e 103 (5.1%)
m/e 166 (22.4%)
m/e 165 (52.8%)

(13)

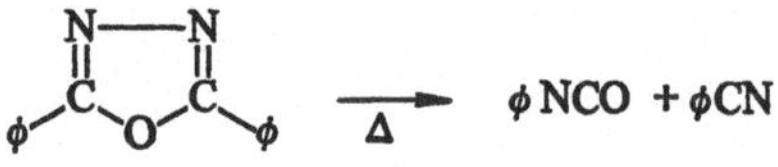
Δ
ϕNCO + ϕCN

(14)

24

Pritchard and Funke have reported the mass spectra and thermal fragmentation of *meso* and *dl*-hydrobenzoin cyclic sulfites, 25.[29] The *meso* and *dl* isomers gave distinct thermal fragmentation patterns (15) but similar mass spectra (16).

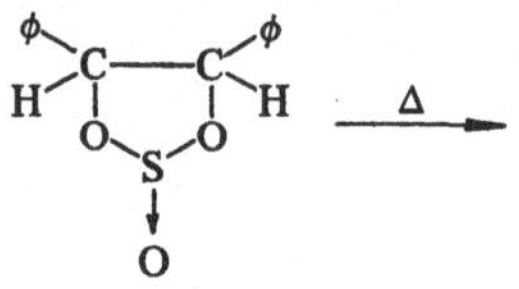

25

Δ	meso (%)	dl (%)
ϕ–CH=CH–ϕ	2	10
ϕ CH₂–C(=O)–ϕ	57	15
ϕ_2 CH–CHO	41	66
other	--	9

(15)

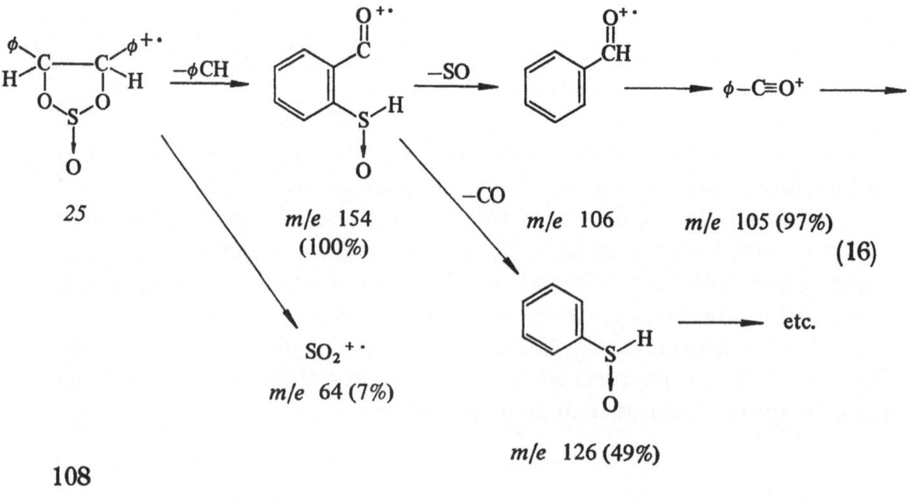
−ϕCH
−SO
ϕ–C≡O+
25
m/e 154
(100%)
m/e 106
m/e 105 (97%)
−CO
SO2+·
m/e 64 (7%)
etc.
m/e 126 (49%)

(16)

The close similarity of the mass spectra of the *meso* and *dl* isomers of *25* suggested [29] that the primary ionic process was ring opening, followed by rearrangement and elimination of benzaldehyde. Electronic excitation of the sulfur chromophore in *25* would probably yield the same general result as the decomposition in a mass spectrometer. Ions corresponding to the thermolysis products, (15), do occur in the mass spectra though their intensity is low, relative abundances being on the order of 1 to 2%. The relative intensities of these ions are quite sensitive to the stereochemistry of the starting material and they generally parallel the changes observed in the thermochemistry.

Jones and Paisley have reported the ionic and thermal fragmentation of 4,6-diphenyl-1,2,3,5-oxathiadiazine-2,2-dioxide, *26*. Thermal decomposition of *26* in the solid state leads to formation of benzonitrile and a brown-colored intractable glass. The major ionic decomposition pathways for *26* are shown in Eq. (17). The failure of 26 to eliminate SO_2 on thermolysis lysis may reflect a heteroatom effect; it may also be due to a medium effect with the SO_2 being trapped before it could escape the solid. It should be noted that the only volatile pyrolysis product, benzonitrile, appears as an intense ion, m/e 103, in the mass spectrum of *26*.

$$\phi CO^+, \phi CON^+, \phi CN_2{}^+$$

$$m/e\ 105, m/e\ 119, m/e\ 117$$

Porter and Spear have reported the mass spectrum of aziridine [30] which shows $HC{\equiv}NH^+$ as the base peak in addition to peaks which correspond to $(M-1)^+$, $(M-2)^{+\cdot}$ and CH_3^+. The mercury sensitized photolysis of aziridine [31] gives predominantly ethylene and nitrene. This heteroatom effect does not apply to the mass spectral photochemical correlation for ethylene oxide, which shows very similar behavior in both systems. [32]

2. Steric Effects

De Jong and Van Fossen have presented one of the clearest examples of a differential steric effect on the reactivity of ions and neutrals. [33] The mass spectrum of the *t*-butyl-*o*-phenylsulfite, *27*, is completely dominated by the $(M-CH_3)^+$ ion at m/e 197, (18). The thermolysis of *27* on the other hand

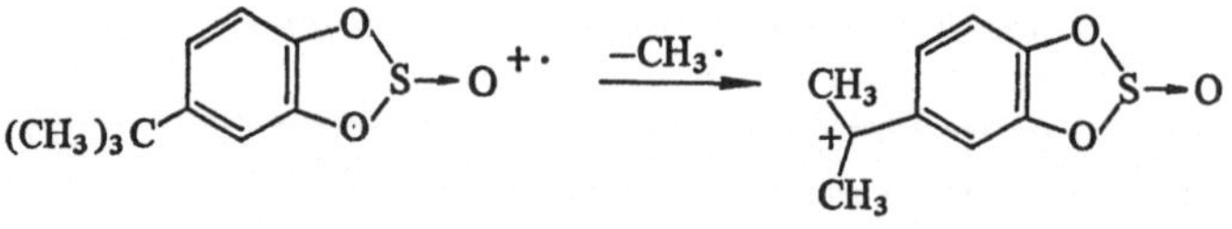

$$\text{(18)}$$

m/e 212

27

m/e 197

results in elimination of CO and SO to give products that arise from *t*-butyl-cyclopentadienione (19).

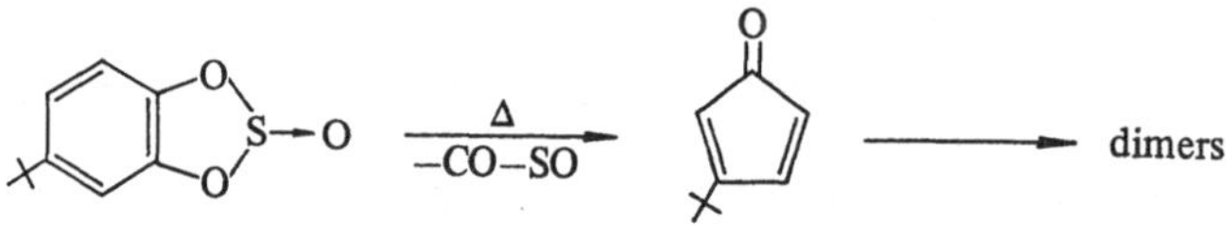

$$\text{(19)}$$

There was a 4% ion in the mass spectrum of *27* that correspond to loss of CO and SO from m/e 197, but the relationship of this ion to the pyrolysis product is too distant to be of any use. The basis for the differential steric effect is most likely a heteroatom effect. Resonance structures like *28* will substantially stabilize the $(M-CH_3)^+$ ion from *27* and they would not stabilize a corresponding pyrolysis product. In the absence of heteroatoms steric effects do not appear to interfere with ionic and thermal fragmentation correlations.[19]

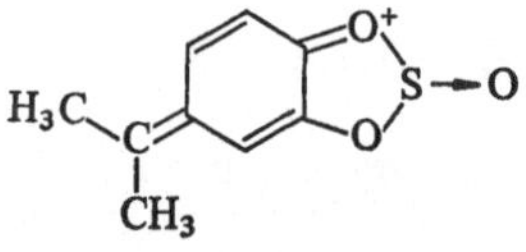

28

Cotter has reported another example of a correlation failure that involves both steric effects and heteroatoms.[34] The mass spectra of the fluorocyclic acetals of formaldehyde, *29*, $n = 1$ to 4, all contained intense ions at m/e 64 ($CF_2=CH_2^{+\cdot}$). When n was 1 or 2 these ions were the base peak. With $n = 3$ or 4, m/e 63 ($FCH_2-\overset{+}{O}=CH_2$) was the base peak. Pyrolysis of all four compounds gave predominantly $CF_2=CH_2$. The formation of m/e 63 in the spectra of *29*, $n = 3$ or 4 is virtually certain to be due to the steric accessibility of the F transfer transition state which gives the heteroatom

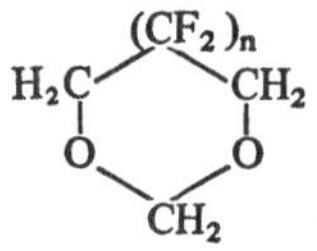

$$(CF_2)_n$$

29

stabilized fragment ion at m/e 63. The absence of analogs of m/e 63 in the thermolysis of *29* is not surprising on the basis of the heteroatom effect.

3. Nonquantitative Correlations

Since reactions in a mass spectrometer are generally less selective than their thermal or photochemical analogs, and since more energy is generally available to reactions of a molecule ion, it is inevitable that some reactions will be more or less unique to mass spectrometry. Whenever this occurs the correlation of ionic reactivity with the reactivity of neutrals will be incomplete.

Wagner has reported the ionic and photochemical fragmentation of a number of amino ketones.[35] On photolysis the ketones *30* and *31*

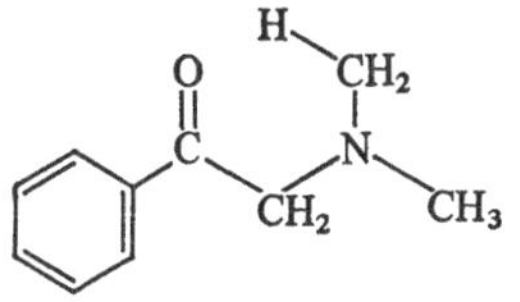

30

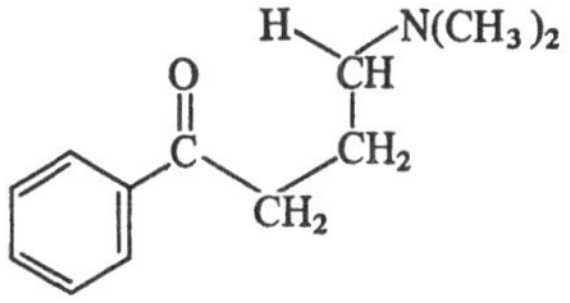

31

gave products related to type II cleavage reactions exclusively. The mass spectra of these compounds had intense ions for γ-hydrogen rearrangement but they also showed ions for α-cleavage at the carbonyl. Thus the correlation of the two types reactivity was not complete, although the α-cleavage ions were genrally less intense than the corresponding McLafferty rearrangement ions. It seems likely that the lack of type I products in the photolysis of *30* and *31* is due to the relatively low energies of the first excited states in these molecules. At higher energies we would expect *30* and *31* to give normal type I cleavage products.

Bursey, Whitten *et al.* have reported a similar example of nonquantitative correlations between mass spectrometry and the photochemistry of ketones.[20] The ketone *32* gives type II cleavage only from its excited singlet state. This reaction has low quantum yield and low relative rate when compared to 2-pentanone. The McLafferty rearrangement in the mass spectra of *32* is always of lower intensity than either the α-cleavage or benzylic cleavage ions; however, the McLafferty rearrangement ion is 25 %

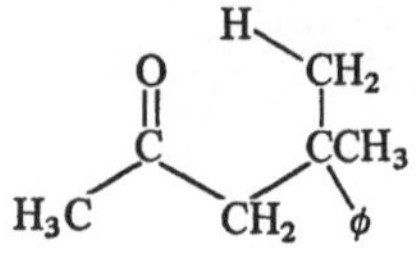

32

the intensity of the benzylic cleavage ion which is the base peak. Again the correlation is not quantitative.

The mass spectra and pyrolysis of alkyl resorcinols are also subject to nonquantitative correlations.[36] Lille and Kundel reported that the mass spectra of 2, 4 and 5 alkyl (C_4 or larger) resorcinols were all dominated by β-cleavage (benzylic) ions. The pyrolysis of the 2 or 4 alkylresorcinols at 700 °C was consistent with their mass spectra; however, pyrolysis of 5-alkylresorcinols gave in addition to the anticipated β-cleavage products, 5,7-dihydroxyindene or an alkenylresorcinol. The yield of the unique olefin was of the same order as the yield of the 5-methylresorcinol. The basis for this reactivity discrepancy may be the scrambling of positions in the resorcinol on electron impact.

Pawda, Bursey *et al.*, have reported the photochemical, thermochemical and mass spectral reactions of a series of N-alkyl-2,4-diphenylpyrrols.[37] The *t*-butyl derivative, *33*, eliminates *iso*-butylene under electron impact, thermal or photochemical excitation. The corresponding cyclohexyl derivative, *34*, eliminates cyclohexene in the mass spectrometer but is relatively

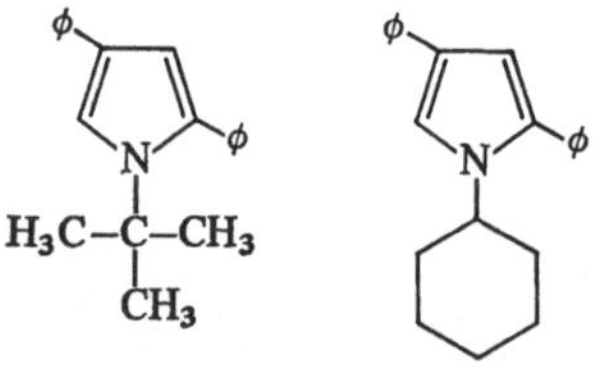

33   *34*

inert to photochemical or thermal excitation. The fact that the appearance potential of the diphenylpyrrole ion from *34* is 2 eV higher than the appearance potential of the corresponding ion from *33* is significant. It is likely that higher energy thermal or photochemical excitation of *34* would also yield the olefin elimination product.

4. Parallels between Ionic Fragments and Neutral Fragments that Differ in Mass by One Dalton

The reactivity of *o*-nitroanisole which was cited above [25] is one example of a reactivity parallel in which the ions in the mass spectrometer have one more proton than the neutrals obtained in thermolysis experiments. Numerous other examples are available in the literature.

Wilson, Barnes and Goldsack have reported the mass spectra and pyrolysis products of a series of nitraminopyridines. The dominant fragmentation patterns of 2-nitraminopyridine under electron impact and thermolysis are shown in Eq. (20) and (21). It appears that the two pathways are directly parallel with the exception of the fact that the radicals that result from the thermolysis pick up a hydrogen atom before they are analyzed.

Closely related cases have been reported by Maccoll et al.[38] and by Shapiro et al.[39] For example the mass spectra of alkyl thiouereas,

$$m/e\ 139 \qquad m/e\ 93 \qquad m/e\ 66 \tag{20}$$

$$\tag{21}$$

R—NH—CS—NHR', are dominated by ions which correspond to RNH^+, $RNHC{\equiv}S^+$, HS^+ and/or RS^+. On thermolysis these same molecules give $RN{=}C{=}S$, $R'NH_2$, $RN{=}C{=}N{-}R'$ and H_2S in direct correspondence (give or take a hydrogen atom) to their mass spectra.

Moller and Pedersen have discussed yet another type of parallel between mass spectral and thermal fragmentation which at first appears to be an exception to the usual correlation.[40] The thermolysis and electronolysis of 35 are illustrated in Eq. (22) and (23) respectively. The absence of thermolysis products at 170 °C which correspond to thiiranes would appear to be an exception to the correlation. It is highly probable in this case that the thiirane was an intermediate in the thermochemical formation of the product

$$\tag{22}$$

$$\tag{23}$$

olefin. The results of the experiment suggest that the thiirane would be desulfurized if subjected to the experimental conditions.

The discussion above clearly illustrated the fact that the correlations between fragmentation in a mass spectrometer and fragmentation in thermolysis or photolysis are not perfect. Indeed no one expected that they would be. The correlations are, however, remarkably close in many cases and an analysis of heteroatom effects and selectivity effects in the three reaction systems should allow one to make numerous useful predictions of reactivity by the simple examination of mass spectra. The next section of this chapter is devoted to a discussion of cases in which the general correlation of reactivity has been a success.

V. Examples of Mass Spectral, Thermochemical and Photochemical Correlations

The following discussion of analogies between the different reactivity systems has been organized according to compound types. The division is merely a matter of convenience and is certainly not exclusive. Generally we will treat thermochemical analogies for a given compound type and then we will discuss the photochemical analogies.

1. Hydrocarbons

The pyrolysis-gas chromatography of a series of aliphatic hydrocarbons has been reported by Fanter and coworkers.[41] Pyrolysis-glc would appear to be a very convenient procedure for establishing reactivity correlations with mass spectra. Unfortunately most of the reactions of simple hydrocarbons under pyrolysis-glc conditions are radical chain reactions. This means that product and fragment ion correspondence should be largely fortuitous in this case. The mass spectra of six C_6H_{12} isomers are substantially similar while the pyrograms of each of the six isomers studied were unique.[41] To be sure, the mass spectra generally contained fragment ions that would correspond to the pyrolysis products, but as we indicated above this correspondence is probably fortuitous. Likewise, one would not expect to find parallels in mass spectra for pyrolytic reactions of aliphatic [42] or aromatic [43] hydrocarbons that lead to the formation of products with molecular weights larger than that of the starting material. Frey and Walch have reviewed the thermal unimolecular reactions of hydrocarbons [44], with particular emphasis on radical cleavage, radical chain processes and reactions that proceed with cyclic transition states, e.g. the Cope rearrangement. The unimolecular thermal fragmentation reactions of hydrocarbons should have analogs in the mass spectra of the same compounds, and they generally do.

114

Swinehart and coworkers and Wilson and coworkers have investigated the pyrolysis of methyl [45] and ethyl [46] cyclobutanes respectively. The primary path for thermal and ionic decomposition of these compounds are shown in Eq. (24) and (25) respectively.

$$
\square^{R} \xrightarrow{\Delta} \| + \|^{R} \tag{24}
$$

$$
\square^{R}_{+\cdot} \xrightarrow{-C_2H_4} {}^{+\cdot}\|^{R} \xrightarrow{\beta \text{ cleavage}} \underset{H_2C}{\overset{C}{\diagup}}\underset{+}{\overset{\diagdown}{C}H_2} \tag{25}
$$

Loudon, Maccoll and Wong reported an extensive investigation of the pyrolytic [47] and mass spectrometric [48] reactions of tetralin and a series of related heterocyclis. They have summarized their findings in a table [47] which is reproduced as Table 1. In spite of the occurrance of substantial heteroatom effects in the thermolysis and ionic fragmentation of the heterocycles the correspondence between the mass spectrometry and thermochemistry of these compounds is quite satisfactory. The exaggerated importance of dehydrogenation products in the thermolysis of tetralin may have been due to radical chain reactions in this system. The retro Diels-Alder fragments in Table 1 correspond to the o-quinodimethane-benzocyclobutene system. Interest in this system prompted the authors to seek other precursors of this hydrocarbon by examining the ionic decomposition of a series of likely starting materials. α-Chloro-o-xylene was chosen because it gave a large $(M-Cl)^+$ ion on electron impact, and indeed on pyrolysis at 630 °C it gave a 70% yield of o-quinodimethane which was isolated as benzocyclobutene or reacted in situ.[47,49]

We have previously mentioned the parallels between the mass spectra and photochemistry of thermochemistry of a series of stilbene-like compounds which were discussed by Johnstone and Ward.[17] The interesting feature of this study was the fact that it was possible to infer the stereochemical course of a number of mass spectral pericyclic reactions on the basis of subsequent fragmentation processes.

The pyrolytic generation of benzyne in parallel to ionic fragmentation reactions has received considerable attention in the literature.[5] One of the procedures that has been used for benzyne generation is the pyrolysis of biphenylene.[50,51] The benzyne ion (m/e 76) has a relative intensity of 25% as compared to the biphenylene molecule ion at 70 eV. Under pyrolytic conditions biphenylene yields a small amount of benzyne although most of

Table 1. A comparison between the electron impact and thermal fragmentation of tetralin and related heterocycles [47,48]

Fragmentation	R. D. A.		M—CH$_3$		M—XH		M—H$_n$(n<2)		M—H$_m$(m>2)		'Styrene'[4]	
Mole % of total products	Th.	E. I.	Th.	E. I.	Th.	I. E.	Th.	E. I.	Th.	E. I.	Th.	E. I.
Tetralin	2	73[1]	20	18[2]	20	18[2]	36	9	25		10	73[1]
Isochroman	80	66						12			11	3
1,2,3,3-Tetrahydro-isoquinoline	66	46		8[5]			7	34	4	13	7	3
Isothiochroman	38	37[6]		14	58	27		7			1	3
Chroman	17	28	39	17		10	24	46			11[6]	
1,2,3,4-Tetrahydro-quinoline	4	3	22	2			36	52	11	19	16	4
Thiochroman	4	25	39	34	23	17	5	9[6]		5[6]	12	10

In all cases except for styrene, the column headings refer to the electron impact fragmentation and the figures for the electron impact case include any subsequent loss of hydrogen (H_2,H) in columns 1, 2, and 3. In the case of column 2 the thermal fragmentation is the loss of methane, column 3 the loss of H_2X and in columns 4 and 5 $n = 2$, $m = 4$ respectively.

In all cases the thermal figures are the result for the lowest % pyrolysis as determined in a flow system.

1) Ref. 48).
2) This is the same fragmentation in this case.
3) Retro-Diels-Alder reaction, in thermal case includes figures for o-xylene produced.
4) Formation of C_8H_8 by reaction other than retro-Diels-Alder, in the thermal case.
5) In this case $(M—CH_4)^+$.
6) No metastable for formation of this ion. X = O, NH or S.

the reaction products result from cleavage of a single bond to give the diradical *36* which subsequently dimerizes or reacts with other substrates in the pyrolysis system.

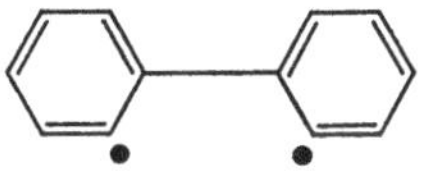

36

Pyrolysis of benzene and its deuterated derivatives has been shown to proceed primarily by formation of phenyl radicals or their equivalent and a small amount of benzyne.[52] This result is in accord with the mass spectrum of benzene [21] which shows relatively intense fragment ions at m/e 77 and 76.

The flash vacuum pyrolysis and the mass spectrometric fragmentation of a series of five methyl substituted phenanthrenes were correlated with each other in a qualitatively predictable way.[19] The major thermochemical products for these compounds generally corresponded to one of the five most intense fragment ions in the compound's electron impact mass spectrum. The importance of products which corresponded to ions with high appearance potentials increased with increasing temperature in the thermolysis system.

The results for thermolysis and electron impact induced fragmentation of 4-methyl-phenanthrene, *37*, are representative of the series studied by Dougherty, Bertorello and Bertorello.[19] The major singly charged ions in the mass spectrum of *37* are illustrated in Eq. (26).

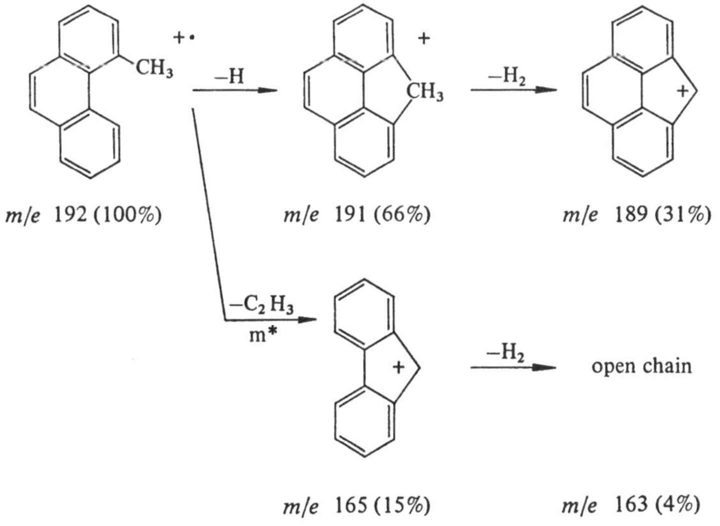

m/e 192 (100%) m/e 191 (66%) m/e 189 (31%)

(26)

m/e 165 (15%) m/e 163 (4%)

117

Pyrolysis of *37* at 750 °C returned 50% of the starting material and gave the products illustrated in Eq. (27). The virtual absence of a peak at m/e 178 in the mass spectrum of *37* which would correspond to the phenanthrene

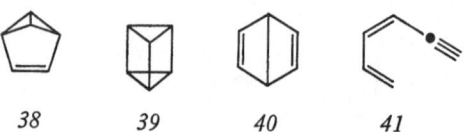

| *37* | 7% | 6% | 25% | (27) |

obtained on pyrolysis suggests that the phenanthrene is not a primary thermolysis product. Phenanthrene was virtually always a product of thermolysis in this series of experiments, and it is likely that it was formed by condensation of the primary products of pyrolysis. One of the interesting observations concerning the methyl phenanthrene series is the fact that the difference in appearance potentials between strained and unstrained analogs in the series correlates quite well with the known strain energies for these compounds.[53] It remains to be seen if this observation is general.

Correlations between the mass spectra and photochemistry of hydrocarbons have generally received less attention than the thermochemical analogies. Nonetheless numerous reactivity correlations can be found in the literature.

The randomization of carbon and hydrogen atoms in benzene which occurs on electron impact [54] is analogous to the numerous skeletal rearrangements in the benzene series that occur upon photolysis.[55,56] The rearrangements of the benzene molecule ion must proceed through structures that are related to known products of benzene photolysis, *i.e.*, benzvalene, *38*, prismane, *39*, Dewar benzene, *40*, and hexadienyne, *41*. The open chain form of the benzene molecule ion, *41*, has been a subject of

| *38* | *39* | *40* | *41* |

discussion among mass spectroscopists for a long time. The isolation of *41* from high energy photolysis of benzene is an interesting example of the general correlation of photochemical reactivity and reactivity in a mass spectrometer.

Cyclic 1,3-diene iron tricarbonyl complexes eliminate hydrogen on electron impact to give predominant odd electron ions with iron bonded to an aromatic system [57] These same molecules eliminate hydrogen and iron on photolysis to give aromatic hydrocarbon products.

2. Alkyl Halides and Halocarbons

The gas phase pyrolysis of alkyl halides has been extensively reviewed [58], and in general the unimolecular gas phase reactions of alkyl halides parallel their reactivity in a mass spectrometer. For example, ethylchloride yields ethylene and HCl on thermolysis [59], and the ethylene ion in the mass spectrum of ethyl chloride is significantly more intense than the molecule ion. 1,2-dichloroethane also eliminated HCl thermolytically and the corresponding ion is the base peak in its mass spectrum. Elimination of HCl is also common to the mass spectra and thermochemistry of chloroprene dimers.[61] Although in this case the major ion at m/e 91 had no definite analog in the thermochemistry. This is probably due to the fact that m/e 91 was a tropylium ion which would not be stabilized as a neutral.

The pericyclic ring opening of halocyclopropanes that occurs on solvolysis was directly paralleled by the gas phase ionic reactivity of the same compounds.[62] The ring opening of _42_ was the most facile in a series (Eq. (28)). The same compound eliminated bromide from the molecule ion much more readily than related isomers. Evidently it makes little difference to the reactivity if the leaving group is Br− or Br ·.

The thermolysis of chlorocyclobutane proceeds as shown in Eq. (29) with the formation of ethylene and butadiene.[63] These same reactions are major processes in the mass spectrum of cyclobutylchloride.

$$ (28) $$

_42_

$$ (29) $$

On pyrolysis chlorobenzene gives phenyl radicals and a small amount of benzyne.[64] The phenyl cation is the most intense fragment in the mass spectrum of chlorobenzene and m/e 76 is the fifth most intense fragment in the spectrum. On photolysis chlorobenzene shows similar reactivity with the primary process being cleavage of the carbon chlorine bond.

3. Nitro Aromatics

The thermochemical reactivity of nitrobenzene is remarkably parallel to its mass spectrum. In a mass spectrometer, nitrobenzene fragments by loss of the NO_2 and also by rearrangement to 43 which gives oxygen containing fragment ions (Eq. (30)).[65,66] The primary products from thermolysis

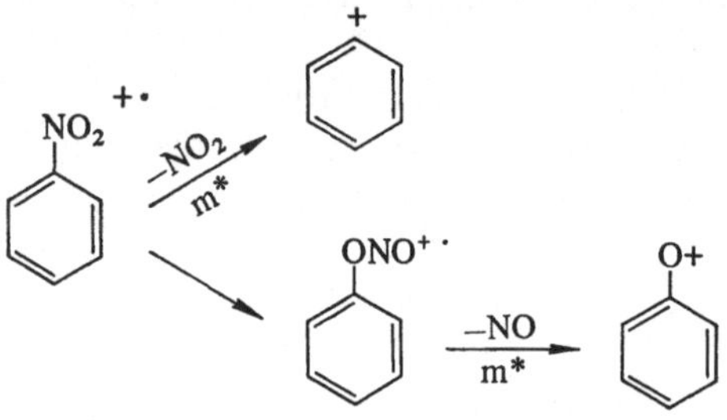

$$\tag{30}$$

43

of nitrobenzene at 600 °C are phenyl radicals and phenoxy radicals. These radicals can either react with the starting material or, as Fields and Meyerson have shown [5,67], the radicals can be used as reagents for reaction with benzene, hexaflurobenzene or other starting materials for polyaryls and arylethers.

The thermal and mass spectral reactions of the dinitrobenzenes are even more interesting.[68] The primary ionic and thermal decomposition reaction is in every case loss of NO_2 to give a nitrophenyl cation or radical respectively. In both systems the nitrophenyl radicals may lose another NO_2 fragment to give phenylene diradicals. In the case of o-dinitrobenzene loss of two NO_2 fragments leads to the formation of benzyne. Benzofurazan, 44, is one of the minor products obtained from thermolysis of o-dinitro-benzene, it is also a minor ion in the electron impact mass spectrum of the compound.

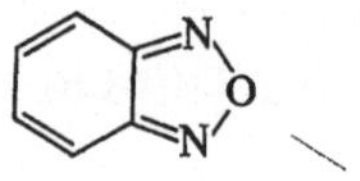

44

Photolysis of nitrobenzene [24] results in formation of phenyl radicals and nitrosobenzene again in direct analogy with the fragmentation of the molecule in a mass spectrometer.

4. Phthalic Anhydride and Related Compounds

Fields and Meyerson were the first to show the utility of phthalic anhydride as a source of benzyne in the gas phase.[5,69] The formation of benzyne from phthalic anhydride, Eq. (31) is directly analogous to the formation of m/e 76 in the mass spectrum of phthalic anhydride, Eq. (32). The benzynes

$$\xrightarrow{690°} \qquad + \text{ CO } + \text{ CO}_2 \qquad\qquad (31)$$

$$\xrightarrow{-\text{CO}_2} \qquad \xrightarrow{-\text{CO}} \qquad\qquad (32)$$

m/e 148 (47%)   *m/e* 104 (100%)   *m/e* 76 (85%)

that are obtained by thermolysis of phthalic anhydride can be used to generate biphenylenes [70], naphthalenes, biphenyls and related hydrocarbons.[5,71]

Heteroatomic dicarboxylic anhydrides also lose CO_2 and CO on electron impact or thermolysis.[72,73,74] In the case of pyrazine-2,3-dicarboxylic anhydride, the initially formed aryne rearranges to a $1:1$ mixture of maleonitrile and fumaronitrile [73], Eq. (33). Quinoxaline-2,3-dicarboxylic an-

$$\xrightarrow[830°]{-\text{CO}_2-\text{CO}} \qquad\qquad (33)$$

hydride gives *o*-phthalonitrile on pyrolysis [74] in direct analogy with the case above. Quinoline-2,3-dicarboxylic anhydride gave *o*-cyano-phenylacetylene on pyrolysis [74], again it appears that the rearrangement of the intermediate isonitrile precludes normal aryne reactivity.

Pyrolysis of 1,2-dicarboxylic anhydrides in the pyrrole series does not lead to the formation of five membered arynes.[75] Anhydrides like *45* do lose CO_2 on pyrolysis but subsequent loss of CO is not competitive with rearrangement or dimerization [75], Eq. (34).

$$(34)$$

The mass spectrum of 4-nitro-phthalic anhydride shows an intense ion at m/e 75 which would correspond to a neutral arynyl radical. The thermolysis of 4-nitrophthalic anhydride gave bimolecular products that could easily be accounted for by assuming the intermediacy of the arynyl radical 46.[76]

$$(35)$$

o-Sulfabenzoic anhydride, 47, also yields benzyne on pyrolysis [70,77], and the benzyne pathway accounts for more than two-thirds of the electron impact induced fragmentation of this compound.

$$(36)$$

Phthalanil, 48, gives large quantities of CO_2 on pyrolysis and electronolysis.[78] The mass spectrum of 48 shows a relatively low intensity

$$(37)$$

ion which would correspond to benzyne; the thermolysis gases from *48* included substantial amounts of CO, suggesting that the benzyne pathway was an important route for the thermal decomposition of *48*.

The mass spectrum of homophthalic anhydride has a base peak at m/e 118 which corresponds directly to benzocyclobuteneone, *49*. The next most intense peak in the spectrum is m/e 90 which corresponds to the elemental composition of fluoenyl allene, *50*.

These two molecules were the major products of hot wire pyrolysis of homophthalic anhydride, Eq. (38).[79,80]

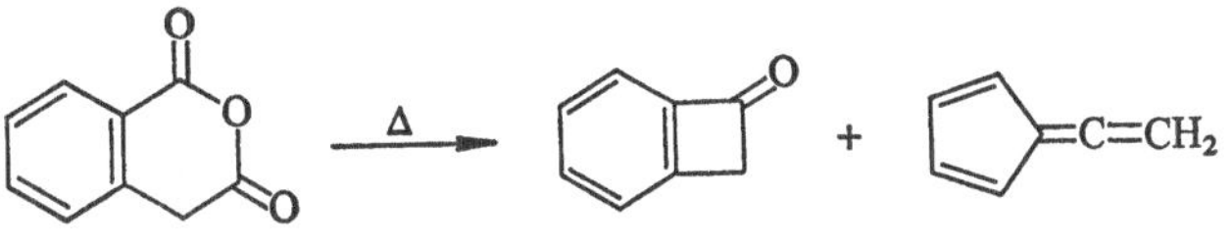

$$ (38) $$

49   *50*

The pyrolytic fragmentation of bicyclo[2.2.1]hepta-2,5-diene-2,3-dicarboxylic anhydride, *51*, is also substantially different from that of phthalic anhydrides [80] and the pyrolysis behavior is closely paralleled in the mass spectrum.

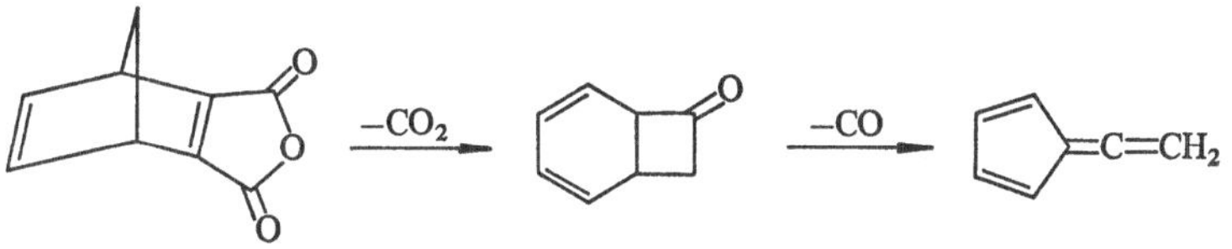

$$ (39) $$

51

The photochemistry of phthalic anhydrides has not received a great deal of attention; however, if reactivity is analogous to other anhydrides [24], photolysis of phthalic anhydride should also provide a convenient route to benzynes.

5. Cyclic Aromatic Sulfites, Carbonates and Related Molecules

o-Phenylene sulfite, *52*, gives cyclopentadieneone on thermolysis (Eq. (40)).[33,81-83] The most intense fragment ion in the mass spectrum of *52*

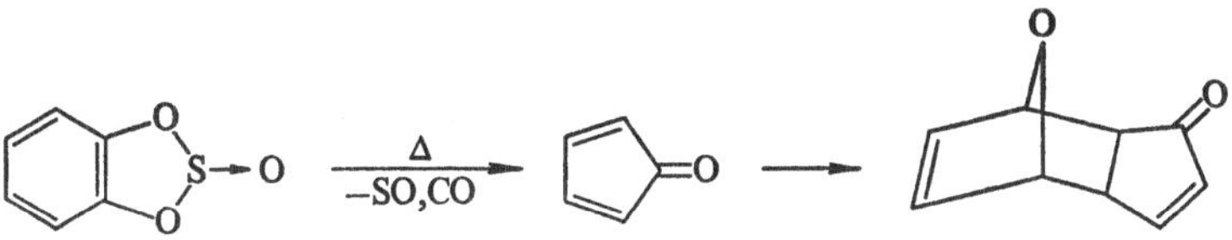

$$ (40) $$

52

occurs at m/e 80 and corresponds to the cyclopentadieneone molecule ion. As we previously mentioned this clear analogy can be distorted by steric and electronic effects in t-butyl substituted systems.[33]

o-Phenylenecarbonate shows remarkably different behavior from the sulfite on both thermolysis and electron impact.[83] In this case the favored reaction path involves loss of CO_2 followed by rearrangement to give the ketene 53. Substituted systems generally follow the patterns established for the parent systems. The pyrolysis and mass spectra of the biphenylene-2,2'-sulfite, 54, also show very close parallels with competitive loss of SO

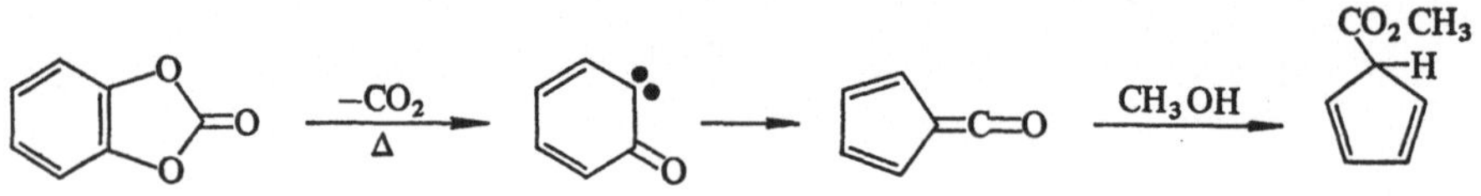

(41)

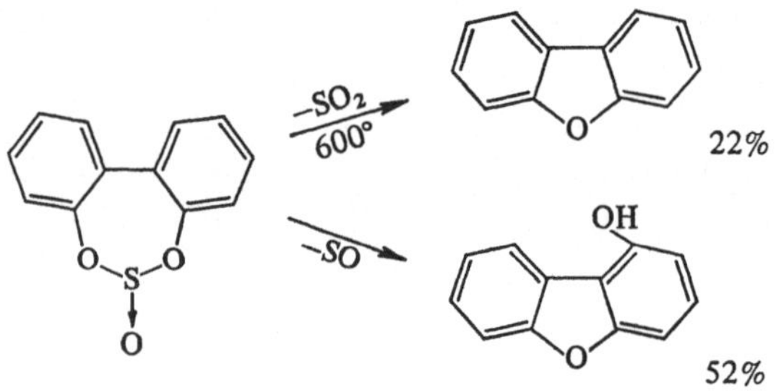

(42)

and SO_2. In this case the reactivity of the carbonate parallels that of the sulfite and thermolysis of the carbonate also results in formation of substantial quantities of dibenzofuran.[81] The cyclic aromatic sulfonic ester, 55, loses SO_2 on both electron impact and thermolysis.[84] Interestingly enough, the oxygen containing intermediate 56 readily traps CO to give high yields of the lactone 57.

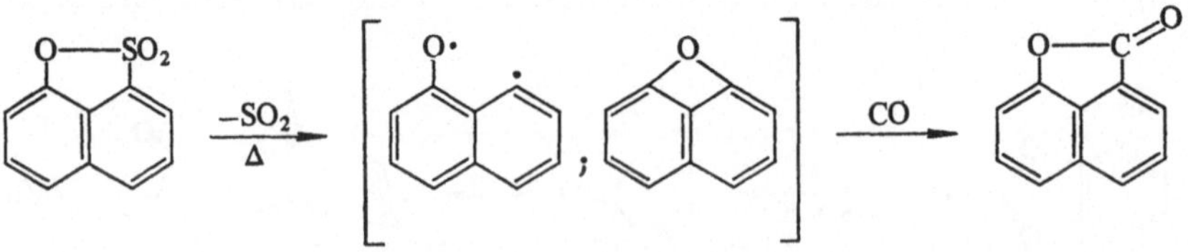

(43)

Oxindole, *58*, is closely related to the cyclic carbonates and sulfites mentioned above. Pyrolysis of *58* and its homologs proceeds primarily by loss of CO followed by rearrangements and bimolecular reactions. [85,86] The pyrolytic and mass spectrometric reactivity are strikingly parallel in this case, in fact the pyrolytic results prompted a reinterpretation of the oxindole mass spectrum. [86]

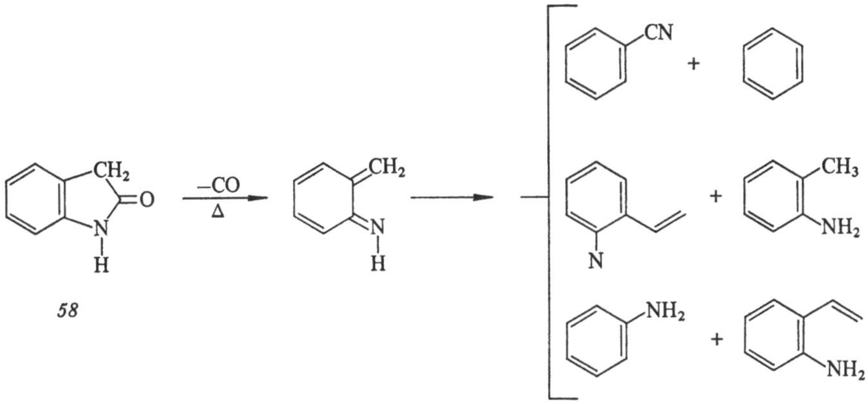

(44)

6. Sulfur Containing Compounds

The Chugaev reaction of S methyl xanthates is a synthetically useful pyrolysiss process. The corresponding electron impact induced elimination is a prominent feature of xanthate mass spectra, (45). [87] The electron impact induced cleavage is at least partially *cis*-stereospecific as the thermal reaction is

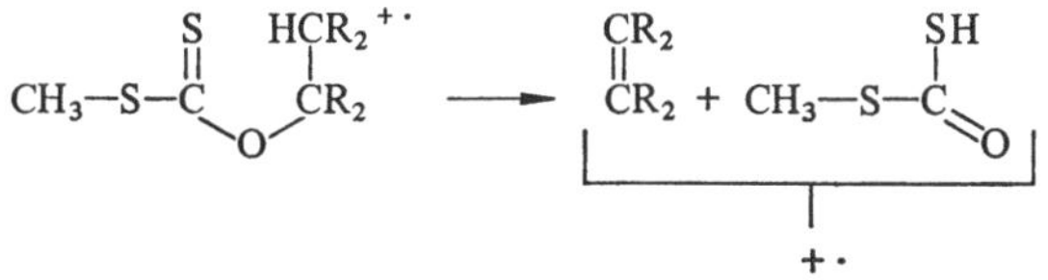

(45)

known to be. The selectivity between secondary and tertiary sites in the pyrolytic and electron impact induced eliminations was qualitatively similar, but the relationship was not quantitative. [87]

Phenyl vinyl sulfides cyclize to substituted benzothiophenes on photolysis. They also rearrange during photolysis to give rearranged benzothiophenes (46). The rearrangement reaction and the elimination of hydrogen (presumably to give the benzothiophene molecule ion) are both prominent processes in the mass spectra of the aryl vinyl sulfides. [88]

$$\text{(46)}$$

Substituted thiophenes are known to rearrange on photolysis [89], very similar processes have been observed to occur in the electron impact induced decomposition of substituted thiophenes.[90] In the mass spectrometer scrambling appears to occur by a reversible ring opening followed by rearrangement.[90] The ring opening has not yet been observed in thiophene photochemical rearrangements; however, it seems likely that it will play some role in the photochemical reactivity of these molecules.

Mercury sensitized photolysis of dimethyl sulfide gives $CH_3S\cdot$ and $CH_3\cdot$ exclusively.[91] The CH_3S^+ ion is the base peak in the corresponding mass spectrum. The dimethyl disulfide system gives $CH_3S_2\cdot$ and $CH_3S\cdot$ radicals on sensitized photolysis.[91] Ions corresponding to both of these products are among the five most intense fragments in the corresponding mass spectrum.

7. Azo Compounds, Oxides, Triazoles and Related Structures

The mass spectrum of diazabasketene, *59*, could have been used to predict the fact *59* would not eliminate nitrogen and give cubane on pyrolysis.

$$\text{+ HCN} \qquad \text{(47)}$$

59

Contrary to intuition, elimination of HCN from *59* is the prominent process thermally [92], and this process could have been reliably predicted from the mass spectrum of *59*.

Azines derived by double condensation of acetylpyridines with hydrazene, decompose primarily by loss of a methyl group in a mass spectrometer.[93] On thermolysis these materials gave complex and more or less intractable tars which would be consistent with decomposition occurring by a similar path.

Azinemonoxides are cleaved by two distinct photochemical pathways. One involves oxygen migration and the other involves a pericyclic ring closure.[94] Both processes are represented in the mass spectra of the compounds. Similar rearrangements occur in both the photochemistry and mass spectrometry of aryl nitrones, aryl N-oxides and aromatic azoxybenzenes.[8]

The primary thermochemical and electron impact induced degradation of aryl azides involves loss of N_2 with formation of an aryl nitrene. The nitrene subsequently reacts in the thermolysis system to give a complex mixture of products (48).[95] An ion corresponding to phenyl nitrene is also prominent

$$\text{(48)}$$

in the mass spectra of oxaziridines and benzotriazoles. Photolysis of 2-phenyloxaziridines results in the formation and subsequent reaction of phenyl nitrene [96], (49).

$$\text{(49)}$$

The mass spectra [97,98], thermochemistry [98–100], and photochemistry [98,101] of benzotriazoles have received considerable attention of late. In virtually every case the primary reaction is loss of nitrogen to give a di-radical which can equilibrate (by rearrangement) with arylnitrene, azepine, and cyanocyclopentadiene structures (50). There are differences between

$$\text{(50)}$$

the photochemical and thermal decomposition of these compounds. The intermediates obtained in the photolysis experiments [101] appear to be substantially more reactive than their thermochemical analogs and reaction with the solvent is a prominent feature of the photochemical decomposition.

J. C. Tov and his coworkers have investigated both the thermochemical and mass spectrometric decomposition of arylarsine derivatives.[102–104] The heterocyclic arsines, 60, decompose both thermally and under electron impact to give dibenzofuran or diphenyleneimine respectively.[102,103] Diphenylarsines also lose the arsenic atom on pyrolysis and electron impact,

$$\text{(51)}$$

60, X = NH or O

but in both cases the elimination is generally preceeded by a pericyclic ring closure followed by elimination of hydrogen.[104]

8. Carbonyl Compounds

Carbonyl compounds constitute a very large class of reactive systems. We have divided the following discussion into thermochemical and photochemical correlations; within each group we will discuss the reactivity of quinones and related structures, alycyclic and open chain carbonyl compounds.

The pyrolysis of indanetrione, 61, yields substantial quantities of benzyne in direct analogy with the mass spectrum of the molecule.[105]

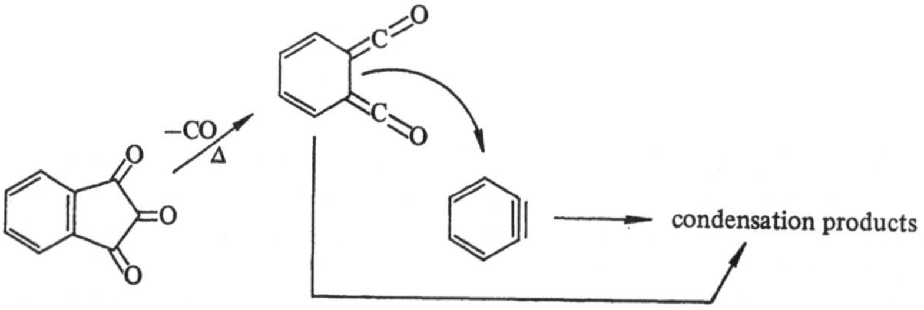

(52)

P-benzoquinone, 62, loses CO in two successive steps under conditions of flash-vacuum thermolysis. The thermal decomposition is precisely analogous to the fragmentation of the molecule ion in a mass spectrometer.[106]

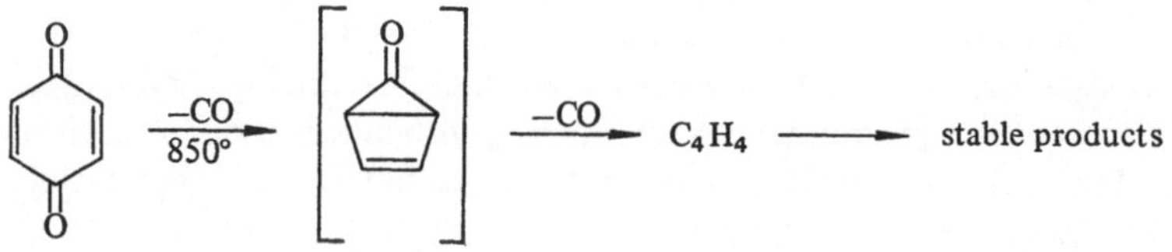

(53)

Pyrolysis of dibenzothiophene-5,5-dioxide, 63, which is closely related to fluorenone, results in the formation of dibenzofuran in direct analogy with the behavior of 63 under electron impact.[107]

The pyrolysis of 1,1,4,4-tetramethyltetralin-2,3-dione, 64, and related compounds gave substantial quantities (ca. 60%) of naphthalene along with small amounts of several related hydrocarbons.[14] The correlation

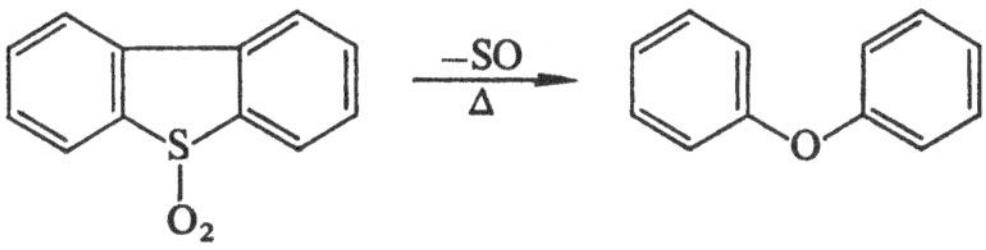

(54)

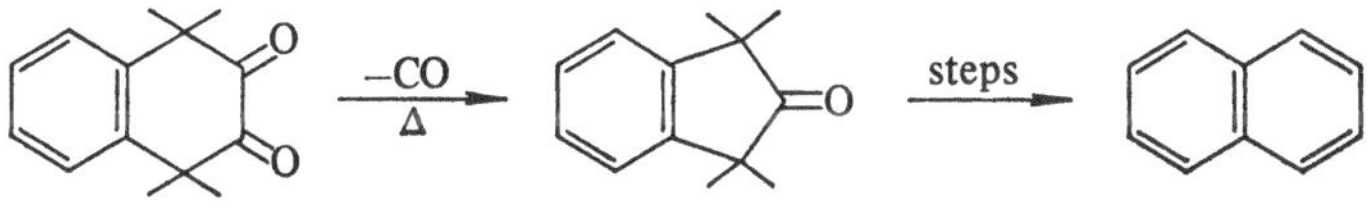

63

(55)

64

of the thermal and electron impact induced fragmentation was quite satisfactory. The reaction sequence to naphthalene from *64* is one of the most complex reaction sequences to be satisfactorily correlated by mass spectra data.[14]

The α-diketones *65—68* all eliminated C_2O_2 on photolysis to give the expected product. Thermolysis results indicated elimination of C_2O_2; however, *65* and *68* gave tars instead of the expected hydrocarbons. The mass spectra of *65—68* showed intense ions for loss of C_2O_2 and low intensity $(M-CO)^+$ ions.[108]

Rees and Yelland have reported a fascinating elimination of CO_2 from nonadjacent carbonyl groups that occurs both on electron impact and thermolysis [109], Eq. (56). Cyclopentanone pyrolysis is substantially complicated by radical chain reactions.[110] Nonetheless the three most abundant pyrolysis products, ethylene, carbon monoxide and 1-butene are all represented in the six most intense fragment ions in the mass spectrum of cyclopentanone.

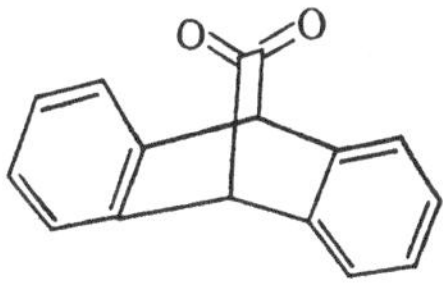

65

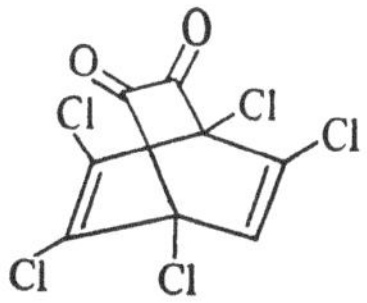

66

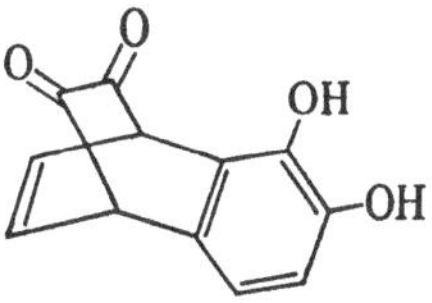

67

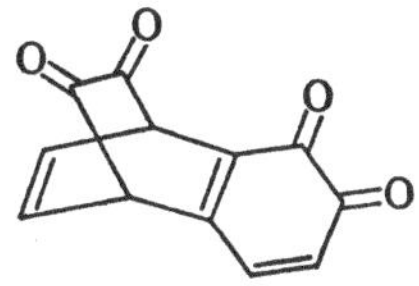

68

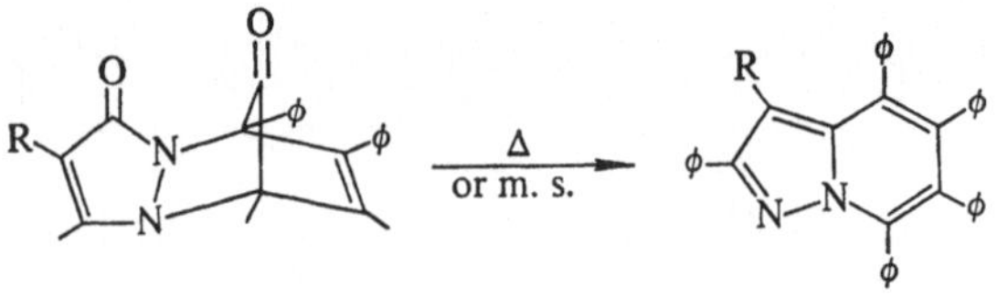

$$\text{(56)}$$

Tetraphenyl cyclopentadieneone decomposes with loss of CO to give "tetraphenylcyclobutadiene" on photolysis, thermolysis and electron impact.[111] By using a direct probe, Brynon, Curtis, and Williams were able to observe strong metastable ions for loss of CO from the parent, and cleavage of the "tetraphenylcyclobutadiene" ion into a charged and an uncharged diphenyl acetylene. Cyclobutadiene itself can be obtained by the flash vacuum pyrolysis of the photo-α-pyrone 69.[112] $C_4H_4^+$· is the base

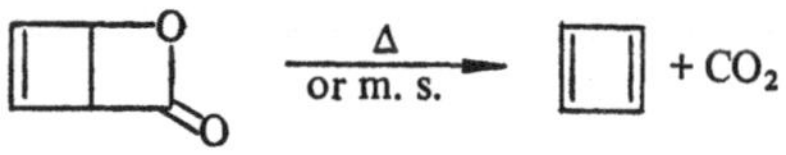

$$\text{(57)}$$

69

peak in the electron impact mass spectrum of 69. Interestingly enough cyclobutadiene is not obtained by pyrolysis of pyrone itself and the m/e 53 ion is of only 0.4 % relative intensity in the α-pyrrone mass spectrum.

Cortisone, 70, and its dihydro derivative both eliminate the elements of acetic acid on thermolysis or electron impact.[113] The similarity between

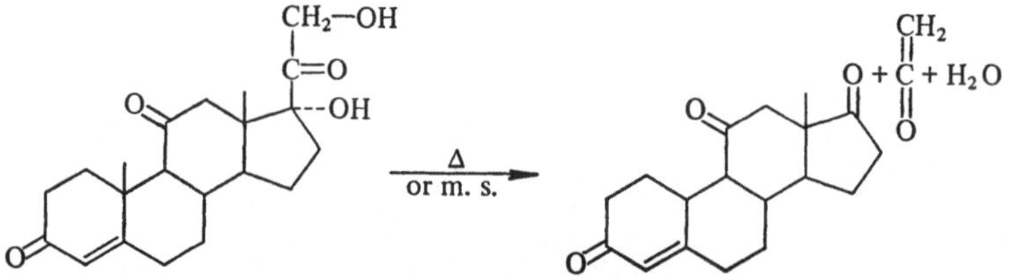

$$\text{(58)}$$

70

the thermochemical and ionic reactivity is so striking in this case that electron impact induced processes have been mistaken for thermochemical ones.

Cava and Spangler have reported that the photochemical, thermo-chemical and electron impact induced decomposition of the diazoketone in Eq. (59) all follow the same pathway [114].

Weiss, Isard and Bonnard have shown that an allylic rearrangement with expulsion of CO_2 occurs in both the mass spectra and thermolysis of benzoate esters of allylic alcohols [115], (Eq. (60)). This pericyclic process

$$\text{(structure)} \xrightarrow[\text{or m. s.}]{\Delta,\, h\nu} \left[\text{(structure)}\right] \xrightarrow{CH_3OH} \text{(structure)} \tag{59}$$

$$\text{(structure)} \xrightarrow[\text{or m. s.}]{\Delta} \text{(structure)} + CO_2 \tag{60}$$

is closely related to the Cope and Claisen rearrangements. It seems very likely that the mass spectra of allylic esters will exhibit rearrangements ions in analogy to the thermal rearrangements in these systems.[116]

Diacetyltartaric anhydride, acetoxy maleic anhydride and diethyl oxaloacetate all yield carbon suboxide, C_3O_2 on pyrolysis. Carbon suboxide and protonated carbon suboxide are prominent ions in the mass spectra of all three of these compounds.[117,118]

The γ-hydrogen rearrangements that are prominent features of the photochemistry and mass spectrometry of many carbonyl systems can also occur with thermochemical activation, Eq. (61) shows one example.[119]

$$\text{(structure)} \xrightarrow[\text{or m. s.}]{\Delta} CH_2=C-C + CH_2 \tag{61}$$

The McLafferty rearrangements is the mass spectrometric analog of the Norrish type II photochemical cleavage of ketones. The relationship between these two processes is easily the most extensively studied of all mass spectral photochemical correlations. In addition to the well known reaction of aldehydes, ketones, and esters, Eq. (62), many analogous γ-hydrogen

$$\text{(structure)} \xrightarrow[\text{or m. s.}]{h\nu} \text{(structure)} + CH_2 \tag{62}$$

rearrangements are known to proceed with either photochemical or electron impact activation, for example see Eqs. (63)[120] and (64).[121]

Several authors have produced experimental [122–124] and theoretical[125] evidence for the step wise nature of the γ-hydrogen rearrangement of

131

$$\text{(63)}$$

$$\text{(64)}$$

molecule ions. This reactivity is directly analogous to the stepwise nature of the Norrish type II elimination.[126,127] Indeed cyclobutanol intermediates have been identified in the electron impact induced decomposition of aldehydes [128] in direct analogy with the competitive photochemical formation of cyclobutanols and ethylenes [127,129,130] Eq. (65).

$$\text{(65)}$$

The photochemical formation of cyclobutanols is substantially favored if the ethylene that would be formed in a type II elimination must have a bridgehead double bond.[130] In 1-adamantylacetone, *71*, the strain is sufficient to substantially inhibit the elimination of acetone from the molecule ion.[130] Cyclobutanol formation is the virtually exclusive photochemical pathway for *71* and related bridgehead acetone derivative.

71

The mass spectra of conjugated cyclohexenones generally show a prominent ion which corresponds to the elimination of ketene from the molecule ion.[131,132] High resolution, metastable ion analysis and isotope labelling were all used to establish the mechanism for ketene elimination as shown in Eq. (66).

$$\text{[structure]} \longrightarrow \text{[structure]} \longrightarrow CH_2=C=O + \text{[structure]}_{+\cdot} \qquad (66)$$

To the extent that the electron impact induced fragmentation with loss of ketene follows the path shown in Eq. (66) the fragmentation is analogous to the rearrangement of conjugated cyclohexenones to bicyclo-[3.1.0]hexan-2-ones. The photochemical and electron impact processes have similar minimal structural requirements of alkyl substitution at carbon 4. Fenselau, Shaffer and Dauben have investigated the related analogy between the photochemical rearrangement (67) and the electron impact induced fragmentation of bicyclo[4.1.0]heptan-2-one, 70 [133], and its homologues. Evidence from fragmentation patterns, and reaction

$$\text{[structure]} \xrightarrow{h\nu} \text{[structure]}_{CH_3} \qquad (67)$$

energetics suggested that the rearrangement did occur in the mass spectrum of 70, and became relatively more important at low ionizing voltages.

Matsuura and Kitaura have reported a semiquantitative correlation between the relative intensities of $M^{+\cdot}$ and $(M-R)^+$ and the ability of systems like 71 to undergo type I photochemical cleavage.

$$\text{[structure]} \xrightarrow[\text{or m. s.}]{h\nu} \text{[structure]} + R\cdot \qquad (68)$$

Turro and his coworkers were among the first to recognize the generality of the correlation between photochemical reactivity and mass spectrometric fragmentation reactions.[135,136] The photochemical reactivity of ketene dimers. Benzylcyclopropane and nortricyclanone [135] as well as pyruvic acid and isopropylpyruvate [136] are substantially reflected in the corresponding mass spectra of these compounds. Furthermore the relative importance of ionic fragmentation pathways, which have photochemical analogs, generally increases at low electron energies. For example at 10 eV the major fragmentation in the mass spectrum of the ketene dimer 72, corresponded to the ketene monomer. This reaction is directly analogous

$$\text{(69)}$$

72

to the photochemical cleavage of ketenes. Pyruvic acid photochemically eliminated CO_2 with a concurrent hydrogen transfer; precisely the same reactivity is shown by the molecule ion.

$$\text{(70)}$$

The photochemical rearrangement of o-nitrobenzaldehyde or its acetals to the corresponding o-nitrosobenzoic acid derivatives has been known for many years. A careful study of the mass spectra and metastable ions obtained from methyl o-nitrosobenzoate and o-nitrobenzaldehyde dimethyl-acetal has provided strong evidence that the electron impact induced fragmentation strongly parallels the photochemical rearrangement [137], Eq. (71).

$$\text{(71)}$$

The photolysis of N-substituted β-keto acid amides results in formation of alkyl isocyanates, Eq. (72).[138] In the mass spectrometer this γ-hydrogen rearrangement increases in relative importance at low electron energies.

Irridation of formanilide or substituted formanilides at 254 nm resulted in quantitative decarbonylation Eq. (73).[139] In the case of formanilide

$$\text{(72)}$$

$$\langle\!\!\!\bigcirc\!\!\!\rangle\text{--NHC--H} \quad \xrightarrow[\text{or m. s.}]{h\nu} \quad \langle\!\!\!\bigcirc\!\!\!\rangle\text{--NH}_2 + \text{CO} \qquad (73)$$

the $(M-CO)^+$ ion was the most intense fragment ion. In the case of m-trifluoromethylformanilide $(M-HCO)^+$ was an intense ion; however, in the case of 3,4,5-trimethoxyformalilide $(M-CO)^+$ ions were not all at prominent in the mass spectrum. This appears to be the result of the very large hetero-atom effect of the three methoxy groups on the ionic fragmentation pathway.

VI. Summary

The unimolecular reactions of ions in a mass spectrometer are remarkably well correlated with both photochemical and thermochemical reactions of the corresponding neutrals. The reactivity correlations are seldom quantitative and exceptions to the correlation should be expected when hetero-atoms or delocalization effects convey unusual stability to the ions in the mass spectrometer as compared to their neutral analogs. In spite of the special effects on ionic stability and the relatively large amount of energy that is available in 70 eV electron impact, it appears that mass spectra will be an increasingly useful guide to new photochemical and thermochemical reactions.

Acknowledgement. The National Science Foundation has generously supported our work

VII. References

1) Stevenson, D. P., Hipple, J. A.: J. Am. Chem. Soc. *64*, 1588 (1942).
2) McLafferty, F. W.: Anal. Chem. *31*, 82 (1969).
3) Meyerson, S.: Rec. Chem. Prog. *26*, 257 (1965).
4) Fields, E. K., Meyerson, S.: Advan. Phys. Org. Chem. *6*, 1 (1968).
5) Fields, E. K., Meyerson, S.: Accounts Chem. Res. *2*, 273 (1969).
6) Maccoll, A.: Modern aspects of mass spectrometry (ed. R. I. Reed), p. 143. New York: Plenum Press 1968.
7) Howe, I.: Mass Spectrometry (ed. D. H. Williams), Vol. 1, p. 82. In: Specialist periodical reports. London: The Chemical Society 1971.
8) Cooks, R. G.: Org. Mass Spectrom. *2*, 481 (1969).
9) Dewar, M. J. S.: The molecular orbital theory of organic chemistry. New York: Mc Graw-Hill 1969.
10) Dewar, M. J. S., Dougherty, R. C.: The PMO theory of organic chemistry. In New York: Plenum, 1974.
11) Dougherty, R. C.: J. Am. Chem. Soc. *90*, 5780, 5788 (1968).
12) Dougherty, R. C.: J. Am. Chem. Soc. *93*, 7187 (1971).
13) Turner, D. W.: Molecular photoelectron spectroscopy. New York: J. Wiley 1970.
14) Brown, R. F. G., Gream, G. E., Peters, D. E., Solly, R. K.: Australian J. Chem. *21*, 2223 (1968).

15) Woodward, R. B., Hoffman, R.: The conservation of orbital symmetry. New York: Academic Press 1970.
16) Smith, E. P., Thornton, E. R.: J. Am. Chem. Soc. 89, 5089 (1967).
17) Johnstone, R. A. W., Ward, D. S.: J. Chem. Soc. C, 1805 (1968).
18) Grubb, H. M., Meyerson, S.: Mass spectrometry of organic ions (ed. F. W. McLafferty p. 516. New York: Academic Press 1963.
19) Dougherty, R. C., Bertorello, H. E., Martinez de Bertorello, M.: Org. Mass Spectrom. 5, 1321 (1971).
20) Bursey, M. M., Whitten, D. G., McCall, M. T., Punch, W. E., Hoffman, M. K., Benezra, S. A.: Org. Mass Spectrom. 4, 157 (1970).
21) Cornu, A., Massot, R.: Compilation of mass spectral data. London: Heyden and Son 1966.
22) Budzikiewicz, H., Djerassi, C., Williams, D. H.: Mass spectrometry of organic compounds. San Francisco: Holden-Day 1967.
23) de Mayo, P., Verdun, D. L.: J. Am. Chem. Soc. 92, 6079 (1970).
24) Calvert, J. G., Pitts, J. N., Jr.: Photochemistry. New York: J. Wiley 1966.
25) Mamer, O. A., Kominar, R. J., Lossing, F. P.: Org. Mass Spectrom. 3, 1411 (1970).
26) Mamer, O. A.: Can. J. Chem. 49, 3602 (1971).
27) Cotter, J. L., Knight, G. K.: Chem. Commun. 1966, 336.
28) Cotter, J. L.: J. Chem. Soc. 1964, 5491.
29) Pritchard, J. G., Funke, P. T.: J. Heterocycl. Chem. 3, 209 (1966).
30) Porter, Q. M., Spear, R. J.: Org. Mass Spectrom. 3, 1259 (1970).
31) Klemm, R. F.: Can. J. Chem., 45, 1693 (1967).
32) Cretanovic, R. F.: Can. J. Chem. 45, 1693 (1967).
33) De Jongh, D. C., Van Fossen, R. Y.: J. Org. Chem. 37, 1129 (1972).
34) Cotter, J. L.: Org. Mass Spectrom 6, 345 (1972).
35) Wagner, P. J.: Org. Mass Spectrom. 3, 1307 (1970).
36) Lille, U., Kundel, H.: J. Chromatogr. 69, 59 (1972).
37) Padwa, A., Gruber, R., Pashayan, M., Bursey, M., Dusold, L.: Tetrahedron Letters 1968, 3659.
38) Baldwin, M. A., Kirkien, A. M., Loudon, A. G., Maccoll, A.: Org. Mass. Spectrom. 4, 81 (1970).
39) Shapiro, R. H., Serum, J. W., Duffield, A. M.: J. Org. Chem. 33, 243 (1968).
40) Møller, J., Pedersen, C. T.: Acta Chem. Scand. 24, 2489 (1970).
41) Fanter, D. L., Walker, J. Q., Wolf, C. J.: Anal. Chem. 40, 2168 (1968).
42) Friedman, M., Bovee, H. H., Miller, S. L.: J. Org. Chem. 35, 3230 (1970).
43) Badger, G. M., Spotswood, T. M.: J. Chem. Soc. 1960, 4420.
44) Frey, H. M., Walsh, R.: Chem. Rev. 69, 103 (1969).
45) Thomas, T. F., Conn, P. J., Swinehart, D. F.: J. Am. Chem. Soc. 91, 7611 (1969).
46) Aspden, J., Khowaja, N. A., Reardon, J., Wilson, D. H.: J. Am. Chem. Soc. 91, 7580 (1969).
47) Loudon, A. G., Maccoll, A., Wong, S. K.: J. Chem. Soc. B, 1733 (1970).
48) Loudon, A. G., Maccoll, A., Wong, S. K.: J. Chem. Soc. B, 1727 (1970).
49) Loudon, A. G., Maccoll, A., Wong, S. K.: J. Am. Chem. Soc. 91, 7577 (1969).
50) Lindow, D. F., Friedman, L.: J. Am. Chem. Soc. 89, 1271 (1967).
51) Friedman, L., Lindow, D. F.: J. Am. Chem. Soc. 90, 2324 (1968).
52) Fields, E. K., Meyerson, S.: J. Am. Chem. Soc. 88, 21 (1966).
53) Jalonen, J., Pihlaja, K.: Org. Mass Spectrom. 6, 1293 (1972).
54) Beynon, J. H., Caprioli, R. M., Perry, W. O., Battinger, W. E.: J. Am. Chem. Soc. 94, 6828 (1972).
55) Wilzback, K. E., Harkness, A. L., Kaplan, L.: J. Am. Chem. Soc. 90, 1116 (1968).
56) Cundall, R. B., Joss, A. J. R.: Chem. Commun. 1968, 902.

57) Dauben, W. G., Lorber, M. E.: Org. Mass Spectrom. *3*, 211 (1970).
58) Maccoll, A.: Chem. Rev. *69*, 33 (1969).
59) Volker, G. W., Heydtemonn, H.: F. Naturforsch. *23*, 1407 (1968).
60) Holbrook, K. A., Walker, R. W., Watson, W. R.: J. Chem. Soc. *B*, 1089 (1968).
61) Brindley, P. B., Nicholson, S. H.: Chem. Ind. (London) *1972*, 118.
62) Baird, M. S., Reese, C. B.: Tetrahedron Letters *1969*, 2117.
63) Cocks, A. T., Frey, H. M.: J. Am. Chem. Soc. *91*, 7583 (1969).
64) Fields, E. K., Meyerson, S.: J. Am. Chem. Soc. *88*, 3388 (1966).
65) Ichimura, T., Mori, Y.: J. Chem. Phys. *58*, 188 (1973).
66) Benoit, F., Holmes, L.: Chem. Commun. *1970*, 1031.
67) Fields, E. K., Meyerson, S.: J. Am. Chem. Soc. *89*, 3224 (1967).
68) Fields, E. K., Meyerson, S.: J. Am. Chem. Soc. *89*, 3224 (1967).
69) Fields, E. K., Meyerson, J.: Chem. Commun. *1965*, 474.
70) Brown, R. F. C., Gardner, D. V., McOmie, J. F. W., Solly, R. K.: Chem. Commun. *1966*, 407.
71) Friedman, L., Lundow, D. F.: J. Am. Chem. Soc. *90*, 2329 (1968).
72) Fields, E. K., Meyerson, S.: J. Org. Chem. *31*, 3307 (1966).
73) Brown, R. F. C., Croward, W. D., Solly, R. K.: Chem. Ind. (London) *1966*, 343.
74) Cava, M. P., Bravo, L.: Chem. Commun. *1968*, 1538.
75) Cava, M. P., Bravo, L.: Tetrahedron Letters *1970*, 4631.
76) Fields, E. K., Meyerson, S.: Tetrahedron Letters *1971*, 719.
77) Meyerson, S., Fields, E. K.: Chem. Commun. *1966*, 275.
78) Cotter, J. L., Dine-Hart, R. A.; Org. Mass Spectrom. *1*, 915 (1968).
79) Spangler, R. J., Kim, J. H.: Tetrahedron Letters *1972*, 1249.
80) Mamer, O. A., Lossing, F. P., Hedaya, E., Kent, M. E.: Can. J. Chem. *48*, 3606 (1970).
81) De Jongh, D. C., Van Fossen, R. Y., Dekovich, A.: Tetrahedron Letters *1970*, 5045.
82) De Jongh, D. C., Van Fossen, R. Y., Bourgeois, C. F.: Tetrahedron Letters *1967*, 271.
83) De Jongh, D. C., Brent, D. A.: J. Org. Chem. *35*, 4204 (1970).
84) De Jongh, D. C., Evenson, G. N.: Tetrahedron Letters *1971*, 4093.
85) Brown, R. F. C., Butcher, M.: Tetrahedron Letters *1970*, 3151.
86) Brown, R. F. C., Butcher, M.: Australian J. Chem. *25*, 149 (1972).
87) Briggs, W. S., Djerassi, C.: J. Org. Chem. *25*, 149 (1972).
88) Weringa, W. D.: Tetrahedron Letters *1969*, 273.
89) Wynberg, H.: Accounts Chem. Res. *4*, 65 (1971).
90) Rennekamp, M. E., Perry, W. O., Cooks, R. G.: J. Am. Chem. Soc. *94*, 4985 (1972).
91) Jones, A., Yamashita, S., Lossing, F. P.: Can. J. Chem. *46*, 833 (1968).
92) McNeil, D. W., Kent, M. E., Hedaya, E., D'Angelo, P. F., Schissel, P. O.: J. Am. Chem. Soc. *93*, 3817 (1971).
93) Buu-Hoi, M. P., Saint-Ruf, G.: Bull. Soc. Chim. France, Ser. *5*, 1861, (1970).
94) Williams, W. M., Dolbier, N. R., Jr.: J. Am. Chem. Soc. *94*, 3955 (1972).
95) Crow, W. D., Wentrap, C.: Tetrahedron Letters *1968*, 5569.
96) Splitter, J. S., Calvin, M.: Tetrahedron Letters *1968*, 1445.
97) Lawrence, R., Waight, E. S.: Org. Mass Spectrom. *3*, 3671 (1970).
98) Ohashi, M., Tseyimoto, K., Yoshino, A., Yonezawa, T.: Org. Mass Spectrom. *4*, 203 (1970).
99) Leonard, N. J., Gdankiewicz, K.: J. Org. Chem. *34*, 259 (1969).
100) a) Wentrap, C., Crow, W. C.: Tetrahedron *26*, 3965 (1970);
     b) Wentrap, C., Crow, W. D.: Tetrahedron *26*, 4915 (1970).
101) Tsujimoto, K., Ohashi, M., Yonezawa, T.: Bull. Chem. Soc. Japan *45*, 515 (1972).

R. C. Dougherty

102) Tov, J. C., Wang, C. S.: J. Organometal. Chem. *34*, 141 (1972).
103) Tov, J. C., Wang, C. S.: Org. Mass Spectrom. *4*, 503 (1970).
104) Tov, J. C., Wang, C. S., Alley, E. G.: Org. Mass Spectrom. *3*, 747 (1970).
105) Brown, R. F. C., Solly, R. K.: Australian J. Chem. *19*, 1045 (1966).
106) Hageman, H. J., Wiersum, U. E.: Tetrahedron Letters *1971*, 4329.
107) Fields, E. K., Meyerson, S.: J. Am. Chem. Soc. *88*, 2836 (1966).
108) Strating, J., Zwanenburg, B., Wagenaar, A., Uccing, A. C.: Tetrahedron Letters *1961*, 125.
109) Rees, C. W., Yelland, M.: Chem. Commun. *1969*, 377.
110) Delles, F. M., Dodd, L. T., Lowden, L. F., Romano, F. J., Duignault, L. G.: J. Am. Chem. Soc. *91*, 7645 (1969).
111) Beynon, J. H., Curtis, R. F., Williams, A. E.: Chem. Commun. *1966*, 237.
112) Hedaya, E., Miller, R. D., McNeil, D. W., D'Angelo, P. F., Schissel, P.: J. Am. Chem. Soc. *91*, 1875 (1969).
113) Adley, T. J., Granata, A.: Org. Mass Spectrom. *5*, 365 (1971).
114) Cava, M. P., Spangler, R. J.: J. Am. Chem. Soc. *89*, 4550 (1967).
115) Weiss, F., Isard, A., Bonnard, G.: Bull. Soc. Chim. France Ser. *5*, (2332), 1965.
116) Lewis, E. S., Hill, J. T., Newman, E. R.: J. Am. Chem. Soc. *90*, 662 (1968).
117) Crombie, L., Gilbert, P. A., Houghton, R. P.: J. Chem. Soc. (*C*), 130 (1968).
118) Crombie, L., Gilbert, P. A., Houghton, R. P.: J. Chem. Soc. (*C*), 137 (1968).
119) Sarner, S. F., Prunder, G. C., Levy, E. J.: Am. Lab., *1971*, 57.
120) Oliver, W. R., Hamilton, L. R.: Tetrahedron Letters *1971*, 1837.
121) Stermitz, F. R., Wei, C. C.: J. Am. Chem. Soc. *91*, 3103 (1969).
122) Briggs, P. R., Shannon, T. W., Vouros, P.: Org. Mass Spectrom. *5*, 545 (1971).
123) Smith, J. S., McLafferty, W.: Org. Mass Spectrom. *5*, 483 (1971).
124) Hass, J. R., Bursey, M. M., Kingston, D. G. I., Tannenbaum, H. P.: J. Am. Chem. Soc. *94*, 5095 (1972).
125) Boer, F. P., Shannon, T. W., McLafferty, F. W.: J. Am. Chem. Soc. *90*, 7239 (1968).
126) Padwa, A., Gergmark, W.: Tetrahedron Letters *1968*, 5795.
127) Lewis, F. D., Turro, N. J.: J. Am. Chem. Soc. *92*, 311 (1970).
128) Fenselau, C., Young, J. L., Meyerson, S., Landis, W. R., Selke, E., Leitch, L. C.: J. Am. Chem. Soc. *91*, 6847 (1969).
129) Yang, N. C., Yang, D. H.: J. Am. Chem. Soc. *80*, 2913 (1958).
130) Sauers, R. R., Gorodetsky, M., Whittle, J. A., Hu, C. K.: J. Am. Chem. Soc. *93*, 5520 (1971).
131) Burlingame, A. L., Fenselau, C., Richter, W. J., Dauben, W. G., Schaffer, G. W., Vietmeyer, N. D.: J. Am. Chem. Soc. *89*, 3346 (1967).
132) Fenselau, C., Dauben, W. G., Schaffer, G. W., Vietmeyer, N. D.: J. Am. Chem. Soc. *91*, 112 (1969).
133) Fenselau, C., Schaffer, G. W., Dauben, W. G.: Org. Mass Spectrom. *3*, 1 (1970).
134) Matsuura, T., Kitaura, Y.: Tetrahedron *25*, 4487 (1969).
135) Turro, M. J., Neckers, D. C., Leermakers, P. A., Seldner, D., D'Angelo, P.: J. Am. Chem. Soc. *87*, 4097 (1965).
136) Turro, N. J., Weiss, D. S., Haddon, W. F., McLafferty, F. W.: J. Am. Chem. Soc. *89*, 3370 (1967).
137) Bursey, M. M.: Tetrahedron Letters *1968*, 981.
138) Reisch, J., Niemeyer, D. H.: Tetrahedron *27*, 4637 (1971).
139) Barnett, B. K., Roberts, T. D.: Chem. Commun. *1972*, 758.

Received March 26, 1973

Chair-Chair Interconversion of Six-Membered Rings

Dr. J. Edgar Anderson

Department of Chemistry, University College, London, England

Contents

I. Introduction

Sachse, at the time in the eighteen-nineties [1] when he first outlined the forms of cyclohexane now known as chair and boat conformations, was aware of the possibilities of interconversion of two mirror-image chair conformations and did consider that the boat conformation might be an intermediate on the interconversion pathway. Mohr [2] in 1918 particularly noted this fact. As interest in conformational analysis developed after 1945 some attention was paid to this particular interconversion and estimates of the potential barrier to the process were made. Thus Shoppee [3] discussing the (chair-boat) interconversion of ring A in androstane, calculated a barrier of 9—10 kcal/mol. It appears from a study of models in conjunction with a reading of the paper that Shoppee was suggesting passage through a cyclohexene like structure *1* as the pathway for interconversion of boat and chair conformations, and that the calculated barrier is in terms of *1*'s being the transition state, but this is not unequivocal. [4]

1   2

Beckett, Pitzer, and Spitzer [6] interpreted infra-red spectral data in terms of an enthalpy of activation for ring inversion of about 14 kcal/mol. [7] These authors excluded a planar conformation as the transition state but specified it no further.

Subsequently, though interest in conformational analysis developed rapidly, there was little mention of the mode of interconversion of six-membered rings until 1960 when the first barrier to such an interconversion was measured by Jensen, Noyce, Sederholm, and Berlin. [8] During this period however interest had been growing in the detailed calculation of energies of conformations by semiempirical methods [9], and these were eventually applied to a range of conformations of cyclohexane including possible transition states for ring interconversion. [10] This led to predictions of the preferred pathway for such a process and the barrier encountered on this pathway.

A general representation of the pathway based on calculations is shown in Fig. 1. Interconversion of two chair conformations *3a* and *3b* requires passing through a high energy conformation *1a* or *2a* to a series of twist-boat and boat conformations *4* and *5*, of intermediate energy. Subsequent passage through other high energy forms such as *1b* or *2b* leads to *3b*. Due to the symmetry of cyclohexane there are several equal-energy conformations of each structure *1,2,4* and *5*. When the ring or the carbon skeleton contains one or more substituents there will again be several conformations of structures *1,2,3,4* and *5* but these will be of different energy.

It is important to recognise that in the case of cyclohexane, and so by extension, of other six-membered rings, the equilibrium ground-state conformation is not an ideal chair with dihedral angles of 60° and carbon-carbon-carbon bond angles of 109.5°.[11] The ring is flattened a little so that dihedral angles are less than 60°, and bond angles are greater than 109.5°. The consequences of this have been much discussed [12], but need not be of great concern here. Both in recent calculations of conformational energies and in discussion within this review, names like "chair", "boat", "twist-boat", or "half-chair" refer not to the highly symmetrical representations which usually correspond to these names, but rather to the minimum energy conformation of structure close to that implied by these highly symmetrical representations.

Most of the values for barriers experimentally measured have been obtained by the nmr technique.[13] Some results have been obtained by ultrasonic methods.[15] It is not intended to discuss the methods used in this review.

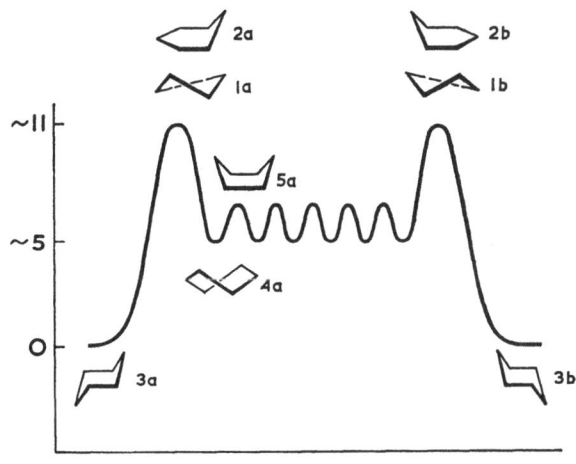

Fig 1.

II. Computations of Conformational Energies

There are generally considered to be three principal contributions [20] to the conformational energy

a) the strain arising from deformation of bond angles away from their preferred lowest energy value (Bayer strain);

b) Torsional strain arising from 1,2 interactions between groups attached to contiguous carbon atoms (Pitzer strain);

c) Interactions between atoms other than those in b), (Van der Waals interactions).

The latter two interactions may occasionally be weakly attractive.

There have been several computations on conformations of cyclohexane which are relevant to the present discussion [10,17a,17b,18a,18b,19,20a,20b] and the results of these are shown in Table 1 as relative enthalpy contributions of total relative enthalpies.

It is not the intention of the present work to give critical comment on computations, since such a review is already available [21], but there are several points which stand out in Table 1. In making them we shall ignore early results [10,17a,20a] when there is subsequent work [19,17b,20b] from the same group. Bond angle strain is agreed to contribute around 2.0 kcal/mol to the barrier to inversion, and to be unimportant in the meta-stable boat and twist-boat conformations. Otherwise in the matter of the relative contributions of torsional strain and of non-bonded interactions, Allinger and his co-workers [17b] by suggesting these are of about equal importance, disagree with Hendrickson [19], with Bucourt and Hainaut [20b], and with Schmid and his collaborators.[18] These latter groups suggest that torsional interactions are much more important than non-bonded interactions, and are alone the source of 70 to 80 percent of the barrier.

In so far as a half chair of type 1 has been postulated as the transition state for the chair-chair interconversion [10], its relative enthalpy represents the barrier to this interconversion, and the values of Bayer strain, Pitzer-strain and van-der-Waals-strain for this form represent the contribution from these factors to the barrier.

While mentioning the transition state it is well to point out that according to Hendrickson [10] it was tacitly assumed before 1961 to be of the form 2.[3-5] In his paper of that year [10] he introduced [3] the half-chair 1, pointing out that it is directly intermediate between the chair 3 and the twist conformation 4, and calculating it to be more stable than form 2 by 1.4 kcal/mol. It has since been generally accepted that this is the transition state.[22] At the same time, as can be seen from Table 1, more recent calculations indicate that the energy difference between 1 and 2 is much less than 1.4 kcal/mol and may be zero [25]. It may be then that 2 is of lower energy than 1, and it would seem that there must be some six-membered rings where this is certainly true, and finally, even if 2 be of greater energy than 1, the difference may be so small that a considerable proportion of the molecules interconverting between chair conformations do so by way of a conformation 2.

Table 1. Calculated enthalpies of conformations[1]) and the contributions to these from various factors

Investigators		Hendrickson		Bucourt and Hainaut		Allinger and Co-workers		Schmid and Co-workers
Reference		10) 1)	19)	20a) 1)	20b)	17a)	17c)	18a)
Date		1961	1967	1965	1967	1967	1971	1967
1 half-chair	B[2])	4.1	2.7		2.1	2.0	1.8	1.9
	P[3])	7.6	7.8		8.2	5.4	4.2	8.4
	VdW[4])	1.0	0.5		0	4.4	4.6	1.0
	Total	12.7	11.0	12.7	10.2	12.0[5])	11.1[6]) (11.04)[7])	11.3
2 half-boat	B	4.5	2.7		2.0			2.0
	P	8.4	7.8		8.2			8.2
	VdW	1.3	1.0		0.1			1.0
	Total	14.1	11.3	14.1	10.4			11.3
4 boat	B	0	0	−0.1	−0.1		1.0	−0.2
	P	5.6	5.4	5.7	5.8		3.2	5.7
	VdW	1.3	1.0	0.3	0.3		2.4	0.8
	Total	6.9	6.4	5.8	6.0	6.4	6.7[6]) (6.6)[7])	6.4
5 twist-boat	B	0	0	0	0		0.7	−0.1
	P	5.1	5.0	5.2	5.2		2.6	5.2
	VdW	0.2	0.7	0.1	−0.1		1.4	0.4
	Total	5.3	5.6	5.3	5.1	5.1	4.8[6]) (4.9)[7])	5.6

1) Values in kcal/mol relative to those calculated for the most stable chairlike conformation except for the results from Ref.[10]) and Ref.[20a]) which are relative to an undistorted chair conformation with tetrahedral bond angles. Small discrepancies in totals may arise from rounding off.

2) Bayer or bond-angle distortion strain.

3) Pitzer or torsional strain.

4) Van der Waals' or non-bonded strain.

5) This total includes a contribution from bond-distortion, see Ref.[17c]).

6) This total includes bond deformation and sketch-bond energy terms[17a]).

7) Results in parentheses are taken from Ref.[17b]); improved versions of these are given in Ref.[17c]).

143

There is no way of determining between this choice, but, where the various contributions to the energies of *1* and *2* have been listed then these contributions are of the same relative importance for *1* and *2*. Thus it is possible to make a useful discussion of barrier heights, though, unaware of the conformation of the transition state. We shall discuss ring inversion in terms of *1* on the understanding that *2* ought also to be considered on similar terms *mutatis mutandi*.

One very important point about the transition state is that it has six kinds of substituent positions (pseudo-equatorial and pseudo-axial at three kinds of carbon atom), compared with only two for the chair. This is important when inversion of substituted cyclohexanes comes to be considered, since there are now several possible pathways each of differing energy. Thus when a single substituent designed to raise the barrier to ring inversion, is introduced, inversion will take place preferentially by way of the transition state of lowest energy, and this may well mean that any constraint due to the substituent is avoided. When R is a bulky group conformation *6* will have particularly high energy due to eclipsing 1,2-interactions but inversion of *7* can occur more readily by way of *8*, where the substituent undergoes fewer interactions.

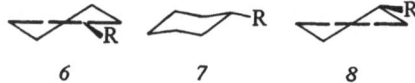

6   7   8

Another way of considering this is that in *7*, at some stage during the inversion, R and adjacent hydrogen atoms must become eclipsed. There is no need for this to happen at the high point of the energy profile in Fig. 1, it can happen at some point in the middle trough in Fig. 1, and the energy of this eclipsing can be considerable and yet not produce an increased barrier to inversion.[26] As a consequence, the barrier to inversion of mono-substituted cyclohexanes has never been found to be sensibly greater than that for cyclohexane itself.

The corollary is also true. If any substitution is made which would tend to lower the barrier to inversion, inversion will prefer to take place by the pathway with the lowest barrier, and a lower barrier will be observed experimentally.

It is worth making another point with reference to *7* where R is a bulky group such as *t*-butyl. Such a molecule exists to a very large extent in the conformation shown rather than in the conformation with the group R axial, so that *7* has been referred to as rigid. There is no evidence that *7* is much more rigid than cyclohexane itself, the barrier to ring inversion appearing to be about the same. What can be said is that the average life-time of a molecule in the conformation 7 is much greater than in the axial conformation.

All calculations agree that the twist-boat conformation 4 is more stable than the true boat 5, but the distinction is small and will be different for different six-membered rings. The meta-stable conformation in the chair-chair inter-conversion pathway is best considered as a rapid boat and twist-boat pseudorotation.[28]

Pickett and Strauss [29,30] have recently calculated the energies of conformations of cyclohexane and some other simple six-membered rings using potential functions derived from vibrational and geometrical data on these molecules. They also quote calculations by Hoffmann [31] using extended Hückel Theory. These calculations do not give a dissection of the various contributions to the barrier and are therefore not relevant to the discussion of experimental results which follows. They suggest the important point that pseudorotation is free in the transition state so that while progress of a given molecule towards the transition state may be well defined the pathway it follows thereafter may not be.

Calculations of the total energies of various conformations on this basis give results in general agreement with earlier calculations. Hoffmann's value [31] for the barrier in cyclohexane 9.9 kcal/mol is the lowest of the calculated values which previously have been higher than the experimentally observed value.

Molecular orbital theory has been applied to the problem of cyclohexane conformations by Dewar and Scholler.[32] Preliminary results using the MINDO/2 method predict a barrier considerably lower than found experimentally.

The inversion of six-membered rings will now be discussed under various subheadings, to illustrate the effect of 1,2-interactions, bond angle strain and van der Waals interactions, but it should be noted that it will not always be possible to separate these effects. Thus replacement of one methylene group by an oxygen atom in cyclohexane will produce different 1,2 interactions along the carbon-oxygen bonds, will change bond angles round the ring, and will remove some hydrogen-hydrogen non-bonded interactions. Thus while a series of heterocyclic compounds may be used to illustrate a particular point, comparisons between heterocyclic and carbocyclic compounds shall be avoided if possible. In a similar way care must be taken in drawing significance from results for highly substituted six-membered rings for the magnitude of barriers may be capable of several different but reasonable explanations, and a choice among those may not be possible. For these reasons, this review is not an exhaustive compilation of barriers to inversion and other information about six-membered rings, but is a selective account of the experimental evidence.

The terms "barriers to ring inversion" shall normally be used when discussing experimental results. By this is meant the free energy of activation for chair-twist [28] interconversion [33] in kcal/mol that is the difference in free-energy between a chair and presumably a half-chair conformation. The free energy of activation for chair-chair interconversion will normally exceed this barrier by 2.3 RT log 2. *i.e.* 1.38 T cal/mol, to allow for the fact that passing on to the inverted chair and return to the original chair are equally likely. On the basis that the entropy of activation is determined only by the symmetry of the ground and transition states, the enthalpy of activation for chair-twist [28] interconversion will normally be equal to this barrier or slightly greater than the barrier by a few hundred calories/mol depending on the individual case. [34]

III. Ring Inversion of Cyclohexane and Deuterated Cyclohexanes

The early nmr work on the inversion of cyclohexane is listed and discussed elsewhere [23]. In 1967 Anet and Bourn [24] reported a study of the ring inversion over an extended temperature range, which should produce thermodynamic parameters for the process more reliable than those previously available, and found a barrier of 10.3 kcal/mol. The enthalpy of activation is 10.8 kcal/mol which is satisfactorily close to the later calculated values (11.0 [19], 10.2 [20b], 11.1 [17c], 11.3 [18a] kcal/mol), while the entropy of activation is +2.8 e. u.

It can be shown that if the entropy of activation be determined by symmetry considerations alone [34] which may be a reasonable assumption, then this should have a value of +3.6 e. u. which is again in good agreement with the experimental value.

These activation parameters are of intrinsic interest and further, their closeness to calculated values of the barrier add interest and credibility to both the total and partial calculated values given in Table 1. Present knowledge of conformations *1,2,4* and *5* is based on these calculations and on a consideration of experimentally measured barriers such as is given in this review.

IV. 1,2-Interactions (Pitzer Strain)

The most striking point to emerge from the calculations is that a large, perhaps predominant part of the barrier to inversion of cyclohexane is due to enhanced 1,2-interactions in the transition state, that is due to the barrier opposing rotation about individual bonds in the ring skeleton. Confirmation that this contribution is important should come from comparisons of barriers to rotation in acyclic systems, and barriers to inversion of the corresponding ring.

In the first instance, only cyclohexane compounds with hydrogen atoms substituted will be considered, so that there should be no drastic alteration of the ring skeleton. In Table 2 are listed values of barriers to ring inversion in cyclohexyl halides where these are known to a reasonable accuracy.[35-37] There is no doubt that there is a significant difference between the values for fluorine and chlorine, but bromine occurs between these two values. The barriers to rotation in the series CH_3CH_2X [38] and in the series $(CH_3)_2CX-C(CH_3)_3$ [39] are also quoted in the table, but the trends there are not reflected in the results for ring inversion.

Table 2. Barriers to ring inversion of cyclohexyl halides, and to rotation in substituted ethyl halides[1])

Substituent	Ring inversion of cyclohexyl halide			Barriers to rotation		Relative values[2])		
	eq-Boat	ax-Boat	A Mean	B CH_3CH_2X	C $(CH_3)_2CX-$ $-C(CH_3)_3$	A′	B′	C′
Reference				Ref. [38]	Ref. [39]			
H [24]	10.3	10.3	10.3	2.88	6.97	0	0	0
F [35] [36]	9.8 9.8	9.6 9.5	9.7 9.65	3.33	8.04	− 0.6	0.5	1.1
Cl [36]	10.7	10.2	10.45	3.68	10.43	0.2	0.8	3.5
Br [37]	10.1	9.6	9.85	3.68	10.73	− 0.5	0.8	3.8

[1]) kcal/mol.
[2]) Relative to the compound with $X = H$, kcal/mol.

It has been shown in the introduction how a single substituent is not expected to raise the barrier to ring inversion. Nor certainly should it lower this barrier at least in terms of Pitzer strain, since it can readily be envisaged that such strain increases in passing from a more or less perfectly staggered chair conformation to the transition state for ring inversion. Thus there is no obvious rationalisation of results quoted in Table 2.

Similar barriers may be expected to obtain with 1,2-disubstituted cyclohexanes, since again these need not be eclipsed in the transition state. This particular point has been made and illustrated by Campbell and Wolfe [27,33,40]. They found that for a series of *cis*-1,2 disubstituted cyclohexanes, the barrier varies between 10.1 and 10.6 kcal/mol where the two substitutents are the same (Cl, OAC, OH, COOM, COOMe). In *trans*-1,2-di-

substituted cyclohexanes the barrier is only 0.2 — 0.3 kcal/mol higher. [27]
In the case of 1,2,4,5-tetrasubstituted cyclohexanes in which there are
necessarily increased eclipsing interactions in the transition state, the bar-
rieres are gratifyingly larger than in cyclohexane (except in the case of
phenyl substituents). These are shown in Table 3 for series 9.[27,33,40]

a) R = OCOCH$_3$

b) R = CH$_3$

9   10

Table 3. Barriers to rotation[1]) in compounds of type 9 as measured by Wolfe and Campbell [27,40]

X	CH$_3$	Cl	N$_3$	OH	OAc	ND$_2$	Ph
Y = COOCH$_3$	11.5	12.8	12.1	11.6	12.5	11.5	10.3
Y = COOH	12.4			11.4			10.4
Y = OCOCH$_3$					12.9		10.9

1) kcal/mol.

It should be noted that the increased eclipsing interactions are steric and
electrostatic in origin since the substituents are more or less polar.

With substitution greater than that in 9 1,2-interactions produce even
higher barriers. A particularly early example was that of the inositol ace-
tates (1,2,3,4,5,6 hexa-acetoxycyclohexane) studies by Brownstein. [41]
The cis-form 10a with all substituents on the same side of the ring has a
barrier to ring inversion at 30 °C of about 15.3 kcal/mol.

A barrier of about 12.8 kcal/mol at — 15 °C has been determined for a
symmetrical 1,2,3,4,5,6-hexachlorocyclohexane [42] while the highest barrier
yet found [43] for a substituted cyclohexane which retains a skeleton with
six carbon atoms is that of the cis form of 1,2,3,4,5,6-hexamethylcyclohexane
10b. The barrier in this instance is 17.0 kcal/mol at 60 °C [45]. It appears [108]
that the barrier to ring inversion in cis-1,2-di-t-butylcyclohexane is 16.3
kcal/mol, a high value that can be explained in terms of the t-butyl groups
being eclipsed or nearly eclipsed in the transition state. The conformational
processes taking place in this molecule are not thoroughly understood.[109]

Another means of studying the contribution of Pitzer strain is to compare the barrier to rotation about bond RA — BR' and the barrier to ring
inversion in $(CH_2)_4 \big\langle^A_B$ where either or both of A and B are heteroatoms,

since it is reasonable to assume that the barrier to rotation about bond A—B in the acyclic case reflects principally 1,2-interactions. This approach has been explored by Harris and Spragg [46] but has not been particularly edifying as will be discussed in Section VIII.

V. Van der Waals' Interactions

The contributions of increased van der Waals' interactions to the barrier to ring inversion has been estimated to be as little as 0.0 kcal/mol [20b] and as much as 4.6 kcal/mol [17c]. When one comes to consider experimental results at present available with this in mind, the difficulty in separating the barrier of inversion into its constituent parts becomes apparent. Since such interactions reflect the bulk of the groups involved, substitution of cyclohexane with groups bigger than hydrogen, in a 1,3, 1,4, or more distant relationship should raise the barrier to ring inversion [47]. Further, from the discussion above, 1,2-interactions associated with the introduction of these substituents will not lower the barrier to inversion though they may not raise the barrier. Thus it would seem that on the basis of Pitzer and van der Waals' strain, the barrier to ring inversion of 1,1-disubstituted [48,49], 1,1,3,3- and 1,1,4,4-tetrasubstituted [48-52] and 1,1,3,5,5-hexasubstituted [51,52] cyclohexanes should be higher than for cyclohexane itself.[53] Experimental results [33] are shown in Table 4 and Table 5 and do not fit this simple hypothesis.

Table 4. Barriers[1]) to ring inversion of $1,1$-di-R_1-$3,3$-di-R_2-$5,5$-di-R_3 cyclohexane

R_1	R_2	R_3	$\Delta G\#$	Ref.
H	H	H	10.3	[24]
CH_3	H	H	10.3	[48]
F	H	H	9.4	[49]
OCH_3	H	H	10.5	[48]
CH_3	CH_3	H	8.7	[50]
CH_3	OCH_3	H	10.2	[48]
OCH_2-	OCH_2-	H	9.3	[48]
CH_3	F	H	9.1	[51]
CH_3	SCH_2-	SCH_2-	9.2	[52]
CH_3	CH_3	F	8.0	[51]
CH_3	CH_3	CH_3	<8	[48]

[1]) kcal/mol.

Table 5. Barriers[1]) to ring inversion of
1,1-di-R_1-4,4-di-R_2 cyclohexane

R_1	R_2	$\Delta G^{\neq}$	Ref.
H	H	10.3	[24)]
CH_3	CH_3	11.4, 11.1	[48,52)]
CH_3	OCH_3	10.9	[48)]
OCH_3	OCH_3	10.7	[48)]
CH_3	F	9.7	[49)]

[1]) kcal/mol.

If the fluoro-compounds be excepted the results in Tables 4 and 5 can be generalised as follows. Barriers for 1,1-disubstituted cyclohexanes are not consistently different from those of the unsubstituted compound while those in 1,1,4,4-tetrasubstituted cyclohexanes are slightly larger than in cyclohexane itself.[43)] Barriers to ring inversion in 1,1,3,3-tetrasubstituted cyclohexanes are slightly lower than in cyclohexane while those of 1,1,3,3,5,5-hexasubstituted cyclohexanes are substantially lower.

These latter two generalisations give no support to the idea that van der Waals' interactions are greater in the transition state but suggest a more obvious explanation. 1,3-syn-diaxial interactions in the ground state may produce a preferred chair conformation that is somewhat flattened.[54)] Since the transition state for ring inversion is undoubtedly flatter than a chair conformation, then in the cases above the ground state conformation is nearer that of the transition state, that is, the barrier is lower. This point seems to be borne out by results for heterocyclic rings given in Section VIII.

The fluoro-compounds show unexpected behaviour, for the barrier in 1,1-difluorocyclohexane is lower than that of cyclohexane which is in turn lower than that of 1,1-dimethylcyclohexane. The barrier in 1,1-dimethyl-4,4 difluorocyclohexane is lower than those of 1,1-dimethylcyclohexane which again in turn is lower than that of 1,1,4,4-tetramethylcyclohexane. The oddity is that in terms of steric size, a fluorine atom is expected to be intermediate *between* a hydrogen atom and a methyl group. It may be that these anomalies and others not so striking are due to fluorine atoms being involved in substantial attractive interactions. It is important that this point be elucidated in view of the widespread use of the fluorine labelling technique in measurement of barriers.[55)]

This discussion may still be oversimplified for we have not made allowance for changes in Bayer strain relative to cyclohexane. This could be a result of interaction of axial substituents which leads to bond angle compression, or of an alteration in the internal bond angles of the ring due to the mutual repulsion of bulky geminal substituents. [56] Unequivocal evidence of the role and relative importance of van der Waals' interactions is quite clearly not available.

VI. Bond Angle Strain (Bayer Strain)

Although all calculations agree that Bayer strain is greater in the transition state for ring inversion than in the ground state, it is difficult to conceive of an unequivocal demonstration of this. Any modification of a six-membered ring to alter bond angle strain is likely to have a much more drastic effect on Pitzer and perhaps on van der Waals' strain.

It is ironic that this should be so since the importance of such strain is particularly easy to visualise. Any molecular models which allow mechanical rotation about carbon-carbon linkages suggest the importance of bond angle strain in inversion of six-membered rings. If a five, seven, eight or greater-membered ring is constructed there is a great deal of flexibility in the model even though bond angles are mechanically constrained to 109.5°. In contrast, the chair conformation of cyclohexane is inflexible, and to invert the ring of a model molecule requires exertion of force. The mechanical constraint of the bond angles to 109.5° has to be opposed to bring about inversion of the ring.

The behaviour of molecular models seems to be reflected in actual molecules as is shown by the results for the barriers to ring inversion of cycloalkanes [24,57-62] shown in Table 6. These results give some indication of the importance of Bayer strain in the case of a six-membered ring. The discussion of spiro-compounds in Section X may also be relevant.

Table 6. Barriers[1]) to ring inversion of cycloalkanes $(CH_2)_n$

n	$\Delta G^{\ddagger}$	Ref.
5	<RT	57)
6	10.3	24)
7	<5.3	59)
8	8.1	60)
9	6	61)
10	5.7	62)

[1]) kcal/mol.

VII. Dipolar Effects

Dipole-dipole interactions play an important role in conformational analysis [63] so it might be expected that a direct effect on barriers to inversion could be demonstrated. We shall show below that much is known about the inversion of heterocyclic six-membered rings, but as was discussed in Section II, when one or more heteroatoms replace carbon atoms in the cyclohexane skeleton large changes in other interactions take place, and may obscure effects due to dipole-dipole interactions. There are some comparisons nonetheless which seem to illustrate the effect of such dipole interactions.

Greenberg and Laszlo [64] have shown that while the barrier in *12* is very similar to that in *11* the barrier in *13* is much lower. The effect of adding a six-membered ring (*11* → *12*) is negligible, but the effect of polar groups in that ring is substantial (*12* → *13*). In a similar way, Brune and

$\Delta G^{\neq} = 10.9$ kcal/mol $\quad \Delta G^{\neq} = 10.9$ kcal/mol $\quad \Delta G^{\neq} = 9.7$ kcal/mol

11 *12* *13*

his collaborators have shown [65] that for *15* and *16*, the reduction in the barrier compared with *14* is much greater in the latter case where the substituent is polar.

14 X = H $\Delta G^{\pm} = 14.9$
15 X = CH$_3$ $\Delta G^{\pm} = 13.8$
16 X = Br $\Delta G^{\pm} = 12.3$

It is not the purpose of this review to explain these results in detail, they are brought forward as clear demonstrations of a polar effect on ring inversion barriers.

VIII. Heterocyclic Six-Membered Rings

We have stated earlier that a knowledge of factors affecting barriers to inversion in cyclohexyl compounds need not allow an explanation of the barrier in a given heterocyclic compound, but it is nevertheless reasonable to hope that comparisons within a series of heterocyclic compounds might point out effects of general applicability.

One interesting effect observed in the heterocyclic series and found to be generally applicable was first noted by Friebolin and his co-workers. [66] They found that while the barrier to inversion of 1,3-dioxan *17* is 9.6 kcal/ mol [67] that in the 2,2-dimethyl derivative *18* is 8.2 kcal/mol. It is expected that the axial methyl group in *18* interact particularly strongly with axial hydrogen atoms in the 4- and 6-positions. [69] In the transition state the direct opposition of these groups may be relieved somewhat [70], so that the barrier to ring inversion is reduced relative to the model compound.

17   *18*

Alternatively, the interaction may produce a flatter chair-conformation which is for that reason more similar to the relatively flat half-chair transition state conformation. Many examples of this effect have been encountered subsequently, and have already been discussed in Section V.

Table 7 shows the barriers to ring inversion of a series of oxygen [67,71,73], [77–79] nitrogen [46,72,74,75,81] and sulphur [67,71,74,76,82] heterocycles. It is as well to separate the *c*-series of compounds with $X = NCH_3$ from that group since comparisons with the *b*-series in the table suggest that the introduction of the methyl group has raised the barrier by more than 2 kcal/mol. This increase might be attributed to increased torsional interactions during inversion, since it is to some extent in line with the difference in rotational barriers between dimethylamine and trimethylamine (about 0.8 kcal/mol) [38] yet it is surprising that for once a simple substitution of one group produces a large increase in the barrier. It might be more reasonable to doubt the validity of neglecting other differences between the series b and c, particularly the effect of nitrogen inversion on ring inversion.

Considering then the series *a*, *b*, and *d* in Table 7 the substitution of one hetero-atom lowers the barrier or has little effect which fits with the mono-substitution effect. The series *19* can be extended to *19e* X = dimethyl-silyl [83] *19f*, X = selenium [71] and *19g*, X = tellurium. [72] For *19e* the barrier to inversion in 5.4 kcal/mol [83] and for *19f* and *19g*, it appears to be much lower than in cyclohexane, or in *19a* to *19d*, since no changes are seen in the nmr spectrum at − 100°. [71] It may be in these cases the low barrier reflects the unusual flatness of the six-membered ring, and perhaps particular ease of rotation about carbon-X bonds, both of which may be consequences of these bonds being relatively long. [84]

Table 7. Barriers to ring inversion[1]) in some heterocyclic six-membered rings

X =		a —O—		b —NH—		c —N(CH$_3$)—		d —S—	
		$\Delta G^{\ddagger}$	Ref.	$\Delta G^{\ddagger}$	Ref.	$\Delta G^{\ddagger}$	Ref.	$\Delta G^{\ddagger}$	Ref.
(structure)	19	9.4	71)	10.1	72)	11.8	72)	8.5	71)
(structure)	20	(12.9)	73)		2)	11.9	75)	11.0, 10.8	74,76)
(structure)	21	9.6	67)		2)	11.3	74)	10.0	67)
(structure)	22	9.4, 9.7	78,79)	10.3	46)	12.5	46)		2)
(structure)	23	10.2	77)		2)	12.2	81)	11.1	82)

[1]) kcal/mol, see f./Ref.[33].
[2]) Not known.

Harris and Spragg [46] have considered the origin of barriers in heterocyclic compounds in some detail and concluded on the basis of known barriers to rotation in acyclic models, that the determining factor in ring inversion barriers is torsional interactions. A consideration of likely transition states and an estimate of the effect of individual heteroatoms allowed them to predict barriers which had not been measured. On this basis the barrier to ring inversion of 1,4-dioxan was predicted to be 9.55 kcal/mol, in very good agreement with the recently measured value of 9.4 — 9.7 kcal/mol. [78,79] On the other hand predicted and measured values for 1,3,5-trioxan are 9.0 kcal/mol and 10.2 kcal/mol [80] respectively, and for 1,3,5-trithian are 7.4 kcal/mol and 11.1 kcal/mol [82] respectively. In these latter cases the discrepancy may arise from the effects of repulsion of 1,3-diaxial lone-pairs.[86]

It is difficult not to conclude however that some result do reflect higher barriers to rotation about single bonds in acyclic compounds. The barriers to rotation in hydrogen peroxide and hydrogen persulphide are 7.0 and 6.8 kcal/mol compared with a value of 2.9 kcal/mol for ethane [38], and these larger barriers seem to be reflected in the ring inversion barriers for the compounds 24—26. [65,87—89]

Barriers have been determined in many other heterocyclic compounds [90] but as can be seen from the discussion of the examples chosen above, interpretation of these results on the basis of individual effects may be difficult.

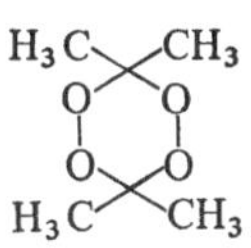

H$_3$C   CH$_3$
O   O
O   O
H$_3$C   CH$_3$

$\Delta G^{\neq}$ 14.9, 15.0 kcal/mol

24

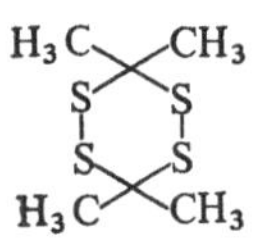

H$_3$C   CH$_3$
S   S
S   S
H$_3$C   CH$_3$

$\Delta G^{\neq}$ 15.6 kcal/mol

25

H$_2$
C
S   S
S   S
S

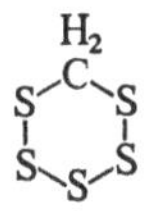

$\Delta G^{\neq} \gg$ 14.0 kcal/mol

26

IX. Spirosubstituted Six-Membered Rings

There have recently been several studies of the inversion of six-membered rings with other rings attached spiro-fashion. [64,91-95]

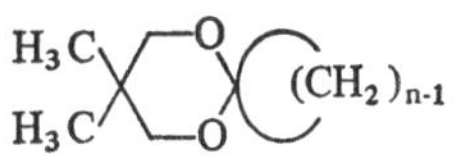

H$_3$C   O
        (CH$_2$)$_{n-1}$
H$_3$C   O

27

Barriers to ring inversion in the series *27* were measured in the hope that as the ring size decreased ($n = 8$ to $n = 4$), the constraint on the bond angle at position 2 of the 1,3-dioxan ring should increase and these barriers should rise. This is borne out by the results [96] shown in the Table 8, but Jones and Ladd [94] have pointed out that an explanation in terms of reduced van der Waals' interactions between the rings is equally reasonable.

Table 8. Barriers to ring inversion[1]) in spiro-compounds [96]

		$\Delta G^{\neq}$	Ref.
27 a	$n = 4$	9.4, 9.3	[91,94]
27 b	$n = 5$	9.0, 8.9	[91,94]
27 c	$n = 6$	8.3, 8.4	[91,94]
27 d	$n = 7$	8.1	[91]
27 e	$n = 8$	8.2	[91]

[1]) kcal/mol.

In the case of a three-membered ring attached spiro-fashion to a six-membered ring, a comparison of the series *28—31* in Table 9 shows that the barrier is reduced in the cyclopropyl compounds. [93,95,96]

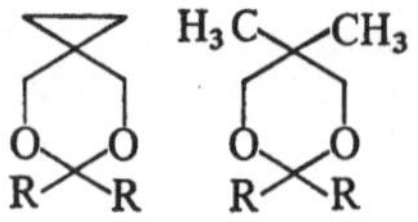

H₃C CH₃ a) R = H
b) R, R = –(CH₂)₃ –
c) R, R = –(CH₂)₄ –

$$28 \qquad 29$$

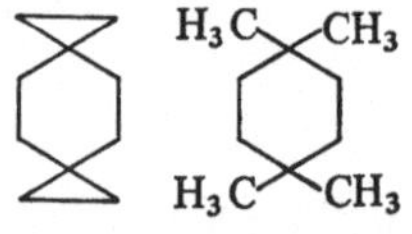

$$30 \qquad 31$$

Table 9. Barriers to ring inversion[1]) in spiro compounds

Compound	28a R = H	28b R = –(CH₂)₃–	28c R = –(CH₂)₄–	30
	9.5	8.9	8.8	10.9
Comparison	29a	29b	29c	31
	10.7	9.4	9.5	11.4, 11.1
Reference	93,96)	93,96)	93,96)	95,48,52)

[1]) kcal/mol.

In the series *29*, van der Waals' interactions of an axial methyl group are considered to be relatively unimportant, but whether these results are then a reflection of bond angle strain is not clear, since it has been pointed out [93] that the barrier to rotation in methyl cyclopropane is substantially lower than that in 2-methylpropane [38]. It can thus be argued that the lower barriers in the compounds *28* and *30* reflect easier rotation about the carbon-cyclopropyl bond.

At a more simple level, on the basis of Lambert's R-value criterion [97], the cyclopropane ring has the effect of flattening the six-membered ring slightly as compared with cyclohexane, and this should manifest itself as a slightly reduced barrier.

There is no unequivocal indication in the spiro-series therefore of an effect of bond angle strain on barriers to ring inversion.

X. Ring Inversion in Fused Six-Membered Rings

Conformational inversion *32* ⇌ *33* of *cis*-decalin must take place by way of a conformation in which both rings are of a boat or twist-boat type, so the high energy point of such an inversion pathway will presumably be some conformation in which one ring is in a boat or twist boat conformation and the other is in the half chair conformation, the usual transition state conformation. At this point the conformational energy of each ring is greater than that of a ground state chair conformation, so that the barrier to inversion in *cis*-decalin should be higher than that of cyclohexane, or than that

32   33

of a *cis*-1,2-dialkyl cyclohexane. This is in fact found, the barrier in *cis*-decalin being 12.9 kcal/mol. [98] Various substituted decalins have also yielded barriers always higher than in corresponding cyclohexanes. [99–102] By contrast the barrier to inversion of the six-membered ring in the perhydroindane series *34* is much lower at about 7 kcal/mol [103]. This is thought to indicate that the five-membered fused ring restricts the flexible transition state less than it does the rigid ground state of the six-membered ring.

34

XI. The Twist Conformation

Until recently knowledge of twist conformation of simple cyclohexanes could be said to be limited to certain di-*t*-butylcyclohexanes which if they existed as chair conformations would have an axial *t*-butyl group, to cyclohexane-1,4-dione and to certain highly substituted cyclohexanes. [104] Undoubtedly twist conformations are quite common, it is unfortunate that reasonably direct evidence is often not available.

Experimental evidence, often an anomalous value for some molecular parameter will indicate that the usual chair conformation is not adopted, but before concluding that a twist conformation is preferred, an explanation in terms of a chair-inverted chair or chair-twist conformational equilibrium must be excluded, since these are more likely, and often this is not possible.

There is however some recent nmr work in which separate signals for chair and for twist type conformations have been observed and which allows a direct measure of their interconversion.

In the nmr spectrum of duplodithiaacetone *35*, separate signals are observed [105] for chair and twist conformations at low temperatures, the twist being the more stable. In the molecule *36* it has been shown [106] that pseudo-rotation within the twist conformation is itself slow an the nmr timescale ($\Delta G^{\ddagger} \sim 10$ kcal/mol).

It would seem that this uncommon behaviour of a six-membered ring is due to the two pairs of sulphur atoms; long sulphur-sulphur and sulphur-carbon bonds make the interactions in a chair conformation very different from those of the cyclohexane chair, so that it would be dangerous to draw conclusions for cyclohexane on the basis of *35* and *36*.

Vinter and Hoffmann have shown [107] that in the bridged compound *37* separate signals are seen for a chair form *37* and a corresponding boat form *38*, the populations being about equal in carbon tetrachloride solution at room temperature, and the barrier to interconversion being about 16 kcal/mol. The constraints placed on the six-membered ring by the two bridges suggests that again this example has little of relevance to simple cyclohexane compounds. Spectral changes for *cis* 1,2-di-*t*-butyl-cyclohexane which have been interpreted in terms of a chair-twist boat equilibrium [108] have since been reinterpreted [109].

35   36   37   38

It has now become clear that in the 1,3-dioxane series *17* certain compounds in which a chair conformation would require 1,3-*syn*-diaxial methyl groups, prefer a twist conformation [110]. Unfortunately the 1,3-interactions which favour the twist conformation also lower barriers to interconversion, so that no direct nmr evidence or barriers have been obtained.

Measurement of the relative energies of twist and chair conformations is possible using ultrasonic relaxation techniques which are suitable for observing equilibria in which there is as little as 0.1% of the higher energy conformer. By combining results from this technique with the earlier nmr results [68] Wyn-Jones and Eccleston [15] have been able to define the conformational system of molecules like 2,2-dimethyl-1,3-dioxan *18* in some detail. At 200 °K the barrier to twist chair interconversion is 4.9 kcal/mol [68], so the twist conformation is thus less stable than the chair by 3.6 kcal/mol, since the chair-twist barrier is 8.5 kcal/mol at this temperature. Results for several other dioxans are also given. When the appropriate nmr information is available, this method allows the determination of the relative energy of unstable boat-type conformations and should extend our knowledge in this relatively unexplored domain.

The subject of twist conformations has recently been reviewed.[128y]

XII. Unsaturated Six-Membered Rings

Endocyclic or exocyclic double bonds in a six-membered ring have the effect of flattening it, and since the transition state for any ring inversion is flatter than the ground state conformation the consequence is that barriers to ring inversion are lower in such rings. In the cyclohexadienes, there is no information on barriers to ring inversion but they are expected to be about 2 or 3 kcal/mol.

In cyclohexene itself, the barrier to ring inversion is certainly low at 5.3 kcal/mol [111,112] and it appears that the preferred conformation is a half-chair *39* in equilibrium with its mirror image conformation *40*. There have been calculations of the energies of cyclohexene conformations by Bucourt and Hainaut [20b] and by Allinger and his colleagues [113] as well as suggestions on the basis of experimental evidence [111,112,114]. These suggestions have been collated in a recent paper [115]. The suggested transition states are *41*, with five atoms coplanar [111], *42* with atoms 2,4,5, and 6 coplanar [113] and the boat conformation *43* [20b,112].

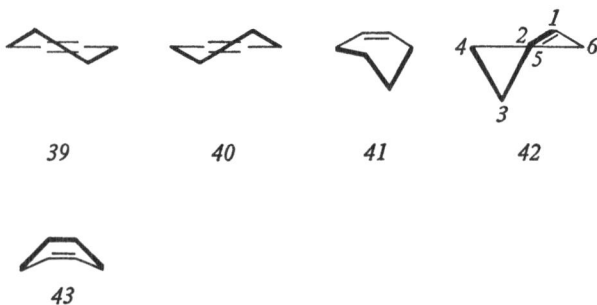

| *39* | *40* | *41* | *42* |

| *43* |

One set of calculations favours *42* [113], the other favours *43* [20b] which is also supported by some experimental evidence [112,116].

It is interesting that while for a monosubstituted cyclohexane there are six possible half-chair transition states, and six different energies corresponding to these, for a mono-substituted cyclohexene with a boat-like transition state *43* there are only two, and in the case of a substituent in the 1-position, only one transition state energies. We mentioned above the absence of a monosubstituent effect on cyclohexane barriers; in contrast it might be expected that monosubstituted cyclohexenes would show enhanced barriers and this has in fact been observed. [36,112,114] There are no reports of cyclohexenes with a substituent in the 1-position, but in this case where the substituent will be eclipsed with the adjacent pseudo-equatorial hydrogen in the 6-position in the transition state it seems reasonable to predict that the barrier will be unusually high.

In the case of cyclohexanone, it now appears that the barrier to ring inversion is less than 5 kcal/mol [117], and so is lower than can be measured by the nmr method at the present time. The barrier has been calculated to be 4.8 kcal/mol [118] and more recently [119] 3.9 kcal/mol by Allinger and coworkers, on the basis that four of the carbon atoms must be coplanar on a transition state quite analogous to the cyclohexane half-chair *1*. Though previous calculations [20b,120] had predicted barriers a little higher than 6 kcal/mol, all agree that the flattening of the ring necessary for ring inversion takes place more readily at the ketone end of the ring. The low barrier is thought to be due to a large extent to the ease of rotation about sp^2—sp^3 bonds [119].

St. Jacques and his co-workers [121,122] have found that barriers to ring inversion in some cyclohexanones substituted with gem-dimethyl groups are higher than in the parent compound, that in 2'2,5,5-tetramethylcyclohexanone being 8.1 kcal/mol [121]. These workers conclude that the enhanced barriers are due to non-bonded interactions of the substituents in the transition state [122], and by a comparison of results with those for methylenecyclohexanes (vide infra) place the barrier to inversion of the cyclohexanone ring itself between 4.5 and 5 kcal/mol.

$$X\diagdown \underset{\|}{C} \diagup X \qquad\qquad X\diagdown \underset{\|}{C} \diagup H$$

43a   *44*

There has been a good deal of work done on methylenecyclohexanes *43* [50,123–126], the barrier in the parent compound *43a* being 8.4 kcal/mol [125,126].

This might be considered a suitable model for studying effects in cyclohexanones, but caution should be exercised since it has been shown [126] that there is a direct relation between barriers in *43* and those for methyl group rotation in *44*. An important fact in *43a* is then steric interactions of the methylene hydrogens, for which there is no counterpart in cyclohexanone. This caution is further justified by the observation [127] of a barrier to chair-chair interconversion in 1,4-*bis*-(methylene) cyclohexane of 7.5 kcal/mol. The corresponding cyclohexan-1,4-dione prefers to adopt a twist-boat type conformation.

XIII. Conclusions

Of the factors concluded from calculations to be important in determining the magnitudes of barriers to inversion of six-membered rings, only that of 1,2 interactions is clearly illustrated by appropriately chosen experimental values. Several compounds with flattened chair conformations due to widely different causes invariably seem to have low barriers to inversion. Several other effects relevant in certain situations can be demonstrated from results for suitable sets of compounds. If a prediction of a barrier be sought, it is my opinion that a better value will be obtained by examining barriers experimentally determined in as similar group of compounds as is available, than by consideration of calculations.

This review does not refer to all barriers to inversion of six-membered rings that have been measured. A more exhaustive compilation is given by Sutherland.[14] As an aid to pursuit of this matter in greater detail, Ref. [128] lists without comment reports not mentioned in this review, or by Sutherland [14], or in Ref. [23], and particularly recent work.

Note Added in Proofs. There has been additional work on the calculation of conformational energies for cyclohexane, see Section II. Wiberg and Boyd [129] conclude that non-bonded interactions make little contribution to the barrier, the most important component of which is torsional strain, in agreement with earlier work. Quantum mechanical calculations of various conformations which do not by their nature allow a splitting of energies into Bayer strain, Pitzer strain, and van der Waals strain, have also been made [130,131]. It has been concluded [131] that transition state conformations *1* and *2* are of similar energies *i.e.* that there is pseudorotation in the transition state as proposed by Pickett and Strauss [29,30].

The most interesting of further experimental results is the barrier to ring inversion of cyclohexanone, see Section XII, of 4.0 kcal/mol [132], in good agreement with the latest calculated value [119]. Further compounds of type *19* have been studied [133,134]. The claims of reference *44* are wrong [135], and have been withdrawn [136]. Vinter and Hoffmann's work [107] on boat conformations has now appeared [137]. There has been further ultrasonic [138] and nmr work [139-141].

Recent work by Lambert and his collaborators [142] on members of series *19* with Group 6 elements shows that the barrier to ring inversion correlates well with torsional barriers in the corresponding CH_3-X-CH_3 compounds. These barriers decrease down the series even though the rings become more puckered, suggesting that bond-angle strain is unimportant, and that torsional strain plays a dominating role. The comments in Section VIII should be modified in the light of these results, at least as far as rings with one heteroatom are concerned.

References

[1] a) Sachse, H.: Ber. Deut. Chem. Ges. *23*, 1363 (1890);
   b) Sachse, H.: Z. Physik. Chem. *10*, 203 (1892);
   c) Sachse, H.: Z. Physik. Chem. *11*, 185 (1893).

[2] Mohr, E.: Prakt. Chem. (2), *98*, 315 (1918).

[3] Shoppee, C. W.: J. Chem. Soc. *1946*, 1138.

[4] In a review of this work written about the same time, Shoppee [5] changes his wording slightly to imply a half boat structure *2* as the transition state for interconversion.

[5] Shoppee, C. W.: Ann. Rep. *43*, 200 (1946).

[6] Beckett, C. W., Pitzer, K. S., Spitzer, R.: J. Am. Chem. Soc. *69*, 2488 (1947).

[7] They were also implicitly aware of an entropy term favouring the transition state and thus implied a free energy of activation for inversion a few hundred calories less than 14 kcal/mol.

[8] Jensen, F. R., Noyce, D. S., Sederholm, C. H., Berlin, A. J.: J. Am. Chem. Soc. *82* 1256 (1960).

[9] For early references see [10].

[10] Hendrickson, J. B.: J. Am. Chem. Soc. *83*, 4537 (1961).

[11] Davis, M., Hassel, O.: Acta. Chem. Scand. *17*, 1181 (1963).

[12] Altona, C., Sundaralingam, M.: Tetrahedron *26*, 925 (1970) and earlier references therein.

[13] For a recent discussion of this technique see Ref. [14] where references to earlier reviews are given.

[14] Sutherland, I. O.: Ann. Rep. N. M. R. Spectr. *4*, 71 (1971).

[15] Eccleston, G., Wyn-Jones, E.: J. Chem. Soc. B *1971*, 2469; and references therein.

[16] Allinger and his coworkers [17a] also concluded that there is a very small contribution (see Table 1) from bond-deformation. Others [20a] have presumed that this contribution is negligible.

[17] a) Allinger, N. L., Miller, M. A., Van Catledge, F. A.: Hirsch, J. A.: J. Am. Chem. Soc. *89*, 4345 (1967);
   b) Allinger, N. L., Hirsch, J. A., Miller, M. A., Tyminski, I. J., Van-Catledge, F. A.: J. Am. Chem. Soc. *90*, 1199 (1968);
   c) Allinger, N. L.: Personal communication, March 1971.

[18] a) Schmid, H. G.: Thesis, Albert-Ludwigs-Universität, Freiburg 1967;
   b) Schmid, H. G., Jaeschke, A., Friebolin, H., Kabuss, S., Mecke, R.: Org. Mag. Res. *1*, 163 (1969).

[19] Hendrickson, J. B.: J. Am. Chem. Soc. *89*, 7036 (1967).

[20] a) Bucourt, R., Hainaut, D.: Bull. Soc. Chim. France *1965*, 1366;
b) Bucourt, R., Hainaut, D.: Bull. Soc. Chim. France *1967*, 4562.

[21] Williams, J. E., Stang, P. J., Schleyer, P. V. R.: Ann. Rev. Phys. Chem. *19*, 531 (1968).

[22] See for example several reviews [14,23,24], and individual papers cited therein.

[23] Anderson, J. E.: Quart. Rev. *19*, 426 (1965).

[24] Anet, F. A. L., Bourn, A. J. R.: J. Am. Chem. Soc. *89*, 760 (1967).

[25] Schmid, and his collaborators [18b] have made the point that there are several conformations similar to the half chair and half boat of comparable energies.

[26] This point has been elegantly demonstrated by Wolfe and Campbell.[27]

[27] Wolfe, S., Campbell, J. R.: Chem. Commun. *1967*, 874.

[28] The term "twist conformation" will be used in this review to represent the pseudo-rotation among boat and twist-boat conformations.

[29] Pickett, H. M., Strauss, H. L.: J. Am. Chem. Soc. *92*, 7281 (1970).

[30] Pickett, H. M., Strauss, H. L.: J. Chem. Phys. *53*, 376 (1970).

[31] Footnote 20 in the Ref. [29], supra.

[32] Unpublished results of Scholler, W. W. quoted by Dewar, M. J. S.: Topics in Curr. Chem./ Fortschr. Chem. Forsch. *23*, 1 (1971).

[33] Literature results have been reduced to the chair → twist barrier value where necessary, though often it is not clear whether results are for chair → chair or for chair → twist interconversion. Where there is any explicit or implicit evidence that a change should be made to give the chair → twist barrier, this has been done. Values calculated at the coalescence temperature are given. In some cases where coalescense of an AB-quartet has been treated as coalescence of a doublet, results have been re-calculated on the basis of an AB quartet. Many results quoted in this review differ from these in the original literature for one or several of these reasons. In some cases barriers reported are for partially deuterated compounds but this may not be pointed out in this text.

[34] Harris, R. K., Sheppard, N.: J. Mol. Spectr. *23*, 231 (1967).

[35] Bovey, F. A., Anderson, E. W., Hood, F. P., Kornegay, R. L.: J. Chem. Phys. *40*, 3099 (1964).

[36] Bushweller, C. H.: Thesis, University of California 1966.

[37] Reisse, J., Stein, M. L., Gilles, J. M., Oth, J. F. M: Tetrahedron Letters *1969*, 1917.

[38] Sources of these values are given by Lowe, J. P.: Prog. Phys. Org. Chem. *6*, 1 (1968).

[39] Anderson, J. E., Pearson, H.: Tetrahedron Letters *1972*, 2779.

[40] Campbell, J. R.: Ph. D. Thesis, Queen's University. Kingston, Ontario, Canada 1969.

[41] Brownstein, S.: Can. J. Chem. *40*, 870 (1962).

[42] Harris, R. K., Sheppard, N.: Mol. Phys. *7*, 595 (1964).

[43] The claim[44] that the barrier to inversion of 1,1-dimethyl-4,4-dibenzyl-cyclohexane is greater than 20 kcal/mol is unconvincing.

[44] Kwart, H., Rank, M. E., Sanchez-Obregon, R., Walls, F.: J. Am. Chem. Soc. *94*, 1759 (1972).

[45] Werner, H., Mann, G., Muhlstadt, M., Kohler, H. J.: Tetrahedron Letters *1970*, 3563.

[46] Harris, R. K., Spragg, R. A.: J. Chem. Soc. B *1968*, 684.

[47] This assumes that at all times van der Waals' forces are repulsive. They may in certain circumstances be attractive but this qualification is relegated to a footnote since any attractive forces involved are expected to be small from the nature of the

J. E. Anderson

Lennard-Jones potential. Further, in the case of bulky substituents such as we are considering here, there are inevitably large van der Waals' repulsive interactions which swamp any small attractive interactions present.

[48] Friebolin, H., Schmid, H., Kabuss, S., Faisst, W.: Org. Mag. Res. *1*, 147 (1969).

[49] Spassov, S. L., Griffith, D. L., Glazer, E. S., Nagarayan, K., Roberts, J. D.: J. Am. Chem. Soc. *89*, 88 (1967).

[50] St. Jacques, M., Bernard, M., Vaziri, C.: Can. J. Chem. *48*, 2386 (1970).

[51] Jefford, C. W., Hill, D. T., Ramey, K. C.: Helv. Chim. Acta *53*, 1184 (1970).

[52] Murray, R. W., Kaplan, M. L.: Tetrahedron *27*, 1575 (1967).

[53] We have chosen these substitution patterns since they must have axial substituents in any chair conformation, and 1,3 diaxial substituent-hydrogen interactions are expected to be the strongest van der Waals' interaction in cyclohexane, We have excluded 1,1,2,2-tetrasubstituted cyclohexanes, since 1,2 interactions are presumably the principal factors in these molecules.

[54] This is a manifestation of the reflex effect, see Sandris, C., Ourissson, G.: Bull. Soc. Chim. France *1958*, 1529; and subsequent references.

[55] Roberts, J. D.: Chem. Brit. *1966*, 529.

[56] The Thorpe-Ingold effect.

[57] This is discussed on pp. 200 ff of Ref.[58]

[58] Eliel, E. L., Allinger, N. L., Angyal, S. J., Morrison, G. A.: Conformational Analysis. New York: Interscience 1965.

[59] Upper limit for cycloheptane deduced from results for 1,1-difluoro-4,4-dimethylcycloheptane given by Glazer, E. S., Ph. D. Thesis, California Institute of Technology, Pasadena 1965.

[60] Anet, F. A. L., St. Jacques, M.: J. Am. Chem. Soc. *88*, 2585 (1966).

[61] Anet, F. A. L., Wagner, J.: J. Am. Chem. Soc. *93*, 5266 (1971).

[62] This result was obtained for 1,1-difluorocyclodecane, Noe, E.A., Roberts, J. D.: J. Am. Chem. Soc. *94*, 2020 (1972).

[63] Riddell, F. G.: Quart. Rev. (London) *21*, 364 (1967).

[64] Greenberg, A., Laszlo, P.: Tetrahedron Letters *1970*, 2641.

[65] Brune, H. A., Wulz, K., Hetz, W.: Tetrahedron *27*, 3629 (1971).

[66] Friebolin, H., Kabuss, S., Maier, W., Lüttringhaus, A.: Tetrahedron Letters *1962*, 683.

[67] Values taken from Ref.[68] which are refinements of those given in Ref.[66].

[68] Friebolin, H., Schmid, H. G., Kabuss, S., Faisst, W.: Org. Mag. Res. *1*, 67 (1969).

[69] This is a consequence principally of carbon-oxygen bonds being shorter than carbon-carbon bonds in cyclohexane.

[70] Since the ultimate effect of a large number of 1,3-diaxial interactions in a six-membered ring is to cause a molecule to adopt a twist conformation, see the section on the twist conformation below, it is reasonable to assume such interactions are considerably relieved in the twist conformation and somewhat relieved in the intermediate half-chair conformation.

[71] Lambert, J. B., Keske, R. G., Weary, D. K.: J. Am. Chem. Soc. *89*, 5921 (1967).

[72] Lambert, J. B., Keske, R. G., Carhart, R. E., Jovanovich, A. P.: J. Am. Chem. Soc. *89*, 3761 (1967).

[73] Calculated from the value for 20d., using a factor derived from tetramethyl analogues[74].

[74] Cleason, G., Androes, G. M., Calvin, M.: J. Am. Chem. Soc. *82*, 4428 (1960); *83* 4357 (1961).

[75] Anderson, J. E.: J. Am. Chem. Soc. *91*, 6374 (1969).

[76] Lüttringhaus, A., Kabuss, S., Maier, W., Friebolin, H.: Z. Naturforsch. *166*, 761 (1961).

164

77) Riddell, F. G.: J. Chem. Soc. B *1967*, 560.

78) Anet, F. A. L., Sandstrom, J.: Chem. Commun. *1971*, 1558.

79) Jensen, F. R., Neese, R. A.: J. Am. Chem. Soc. *92*, 6329 (1971).

80) Pedersen, B., Schlaug, J.: Acta Chem. Scand. *22*, 1705 (1968).

81) Lehn, J. M., Riddell, F. G., Price, B. J., Sutherland, I. O.: J. Chem. Soc. B *1967*, 387.

82) Anderson, J. E.: J. Chem. Soc. B *1971*, 2030.

83) Bushweller, C. H., O'Neill, J. W., Bilofsky, H. S.: Tetrahedron *27*, 3065 (1971).

84) The barriers to methyl group rotation in tetramethyl silane and dimethylselenane are 1.4 and 1.5 kcal/mol respectively[38]. Bond lengths are carbon-silicon, 1.94, carbon-selenium 1.94, carbon-tellurium 2.14; based on the sum of covalent radii[85].

85) Pauling, L.: The nature of the chemical bond, 3rd edit., p. 225. Ithaca, N. Y.: Cornell University Press 1960.

86) Hutchins, R. O., Kopp, L., Eliel, E. L.: J. Am. Chem. Soc. *90*, 7174 (1968).

87) Murray, R. W., Story, P. R., Kaplan, M. L.: J. Am. Chem. Soc. *88*, 526 (1966).

88) Bushweller, C. H., Golini, J., Rao, G. U., O'Neill, J. W.: J. Am. Chem. Soc. *92*, 3055 (1970).

89) Feher, F., Degan, B., Söhngen, B.: Angew. Chem. Intern. Ed. Engl. *7*, 301 (1968). Lower limit for barrier on basis of data given.

90) A recent listing of these can be found in Ref.[14].

91) Anderson, J. E.: Chem. Commun. *1969*, 669.

92) Orohovats, A. S., Dimitrov, V. S., Spassov, S. L.: J. Mol. Struct. *6*, 405 (1970).

93) Anderson, J. E.: Chem. Commun. *1970*, 417.

94) Jones, V. I. P., Ladd, J. A.: Trans. Faraday Soc. *66*, 2998 (1970).

95) Lambert, J. B., Gosnell, Jr., J. L., Bailey, D. S., Greifenstein, L. G.: Chem. Commun. 1004 (1970). It is not clear here whether results are for chair-boat or chair-chair interconversion.

96) The results in Tables 8 and 9 are slightly different from those in references [91] and [93] due to a correction in temperature measurement.

97) Lambert, J. B.: J. Am. Chem. Soc. *89*, 1836 (1967).

98) Jensen, F. R., Beck, B. H.: Tetrahedron Letters *1966*, 4523.

99) Riddell, F. G., Robinson, M. J. T.: Chem. Commun. *1965*, 227.

100) Gerig, J. T., Roberts, J. D.: J. Am. Chem. Soc. *88*, 2791 (1966).

101) Geens, A., Tavernier, D., Anteunis, M.: Chem. Commun. *1967*, 1088.

102) Altman, A., Gilboa, H., Ginsburg, D., Loewenstein, A.: Tetrahedron Letters *1967*, 1329)

103) Lack, R. E., Roberts, J. D.: J. Am. Chem. Soc. *90*, 6997 (1968).

104) For a review see p. 469 of Ref.[58].

105) Bushweller, C. H., Golini, J., Rao, G. U., O'Neill, J. W.: J. Am. Chem. Soc. *92*, 3055 (1970) and earlier work cited therein.

106) Bushweller, C. H., Rao, G. U., Bissett, J. H.: J. Am. Chem. Soc. *93*, 3058 (1971).

107) Vintner, J. G., Hoffmann, H. M. R.: Private communication.

108) Kessler, H., Gusowsky, V., Hanack, M.: Tetrahedron Letters *1968*, 4665.

109) Riecker, A., Kessler, H.: Tetrahedron Letters *1969*, 1227; see footnote 10.

110) a) Thermochemical data, Pihlaja, K., Luoma, S.: Acta Chem. Scand. *22*, 2401 (1968);

b) ^{1}H nuclear magnetic resonance, Nader, F. W., Eliel, E. L.: J. Am. Chem. Soc. *92*, 3050 (1970);

c) optical rotation measurements, Tocanne, J. F.: Bull. Soc. Chim. France *1970*, 750;

d) ebullioscopic data, Kellie, G. M., Riddell, F. G.: Chem. Commun. *1972*, 42;

e) ^{13}C nuclear magnetic resonance, Kellie, G. M., Riddell, F. G. J. Chem. Soc. B. *1971*, 1030.

[111] Anet, F. A. L., Haq, M. Z.: J. Am. Chem. Soc. *87*, 3147 (1965).

[112] Jensen, F. R., Bushweller, C. H.: J. Am. Chem. Soc. *87*, 3285 (1965).

[113] Allinger, N. L., Hirsch, J. A., Miller, M. A., Tyminski, I. J.: J. Am. Chem. Soc. *90*, 5773 (1968).

[114] Jensen, F. R., Bushweller, C. H.: J. Am. Chem. Soc. *91*, 5774 (1969).

[115] Anderson, J. E., Roberts, J. D.: J. Am. Chem. Soc. *92*, 97 (1970).

[116] Gilboa, H., Altman, J. E., Loewenstein, A.: J. Am. Chem. Soc. *91*, 6062 (1969).

[117] Jensen, F. R., Beck, B. H.: J. Am. Chem. Soc. *90*, 1066 (1968).

[118] Allinger, N L., Hirsch, J. A., Miller, M. A., Tyminsky, I. J.: J. Am. Chem. Soc. *91*, 337 (1969).

[119] Allinger, N. L., Tribble, M. T., Miller, M. A.: Tetrahedron *28*, 1173 (1972).

[120] Allinger, N. L., Allinger, J., Da Rooge, M. A.: J. Am. Chem. Soc. *86*, 4061 (1964).

[121] Bernard, M., St. Jacques, M.: Chem. Commun. *1970*, 1097.

[122] St. Jacques, M., Bernard, M., Canuel, L.: Results presented at 2nd International Symposium on Nuclear Magnetic Resonance, Guildford, United Kingdom, July 1972.

[123] Gerig, J. T.: J. Am. Chem. Soc. *90*, 1065 (1968).

[124] Jensen, F. R., Beck, B. H.: J. Am. Chem. Soc. *90*, 1066 (1968).

[125] Gerig, J. T., Rimmerman, R. A.: J. Am. Chem. Soc. *92*, 1219 (1970).

[126] Bernard, M. St. Jacques, M.: Can. J. Chem. *48*, 3039 (1970).

[127] St. Jacques, M., Bernard, M.: Can. J. Chem. *47*, 2911 (1969).

[128] Preliminary communications, the full report of which is referred to, are not included.

a) Yamaguchi, S., Brownstein, S.: J. Phys. Chem. *68*, 1572 (1964);

b) Corfield, G. C., Crawshaw, A.: J. Chem. Soc. B *1969*, 495;

c) Jensen, F. R., Bushweller, C. H.: J. Am. Chem. Soc. *88*, 4279 (1966);

d) Abraham, R. J., MacDonald, D. B.: Chem. Commun. *1966*, 188;

e) Dalling, D. K., Grant, D. M., Johnson, L. F.: J. Am. Chem. Soc. *93*, 3678 (1971);

f) de Jongh, F., Wynberg, H.: Tetrahedron *22*, 583 (1966);

g) Reusch, W., Anderson, D. F.: Tetrahedron *22*, 583 (1966);

h) Gerig, J. T., Ortiz, C. E.: J. Am. Chem. Soc. *92*, 7121 (1970);

i) Ellet, J. D. Jr., Haeberlin, G., Waugh, J. S.: J. Am. Chem. Soc. *92*, 411 (1970);

j) Atalla, R. H.: Spectrochim. Acta *25A*, 889 (1969);

k) Bhacca, N. S., Horton, D.: J. Am. Chem. Soc. *89*, 5993 (1967);

l) Durette, P. L., Horton, D.: Chem. Commun. *1969*, 516;

m) Kalff, M. T., Havinga, E.: Rec. Trav. Chim. *85*, 467 (1966);

n) Angiolini, L., Jones, R. A. Y., Katritzky, A. R.: Tetrahedron Letters *1971*, 2209;

o) Kopf, H., Block, B., Schmidt, M.: Ber. *101*, 272 (1968);

p) Lambert, J. B., Mixan, C. E., Bailey, D. S.: J. Am. Chem. Soc. *94*, 208 (1972);

q) Lee, J., Orrell, K. G.: Trans Faraday Soc. *63*, 16 (1967);

r) Lett, R. G., Petrakis, L., Ellis, A. F., Jensen, A. K.: J. Phys. Chem. *74*, 2816 (1970);

s) Murray, R. W., Kaplan, M. L.: Tetrahedron *25*, 1651 (1969);

t) Riddell, F. G., Lehn, J. M.: J. Chem. Soc. B. *1968*, 1224;

u) Schacht, R. J., Rinehart, K. L.: J. Am. Chem. Soc. *89*, 2239 (1967);

v) Wood, G., McIntosh, J. M., Miskow, M.: Tetrahedron Letters *1970*, 4895;

w) Wood, G., Srivastava, R. M.: Tetrahedron Letters *1971*, 2937;

166

x) Schneider, H.-J., Price, R., Keller, T.: Angew. Chem. Intern. Ed. Engl. *10*, 730 (1971);

y) Kellie, G. M., Riddell, F. G.: Topics Stereochem. in the press.

129) Wiberg, K. B., Boyd, R. H.: J. Am. Chem. Soc. *94*, 8426 (1972).

130) Hoyland, J. R.: J. Chem. Phys. *50*, 2774 (1969).

131) Komornicki, A., McIver, J. W.: J. Am. Chem. Soc. *95*, 4512 (1973).

132) Anet, F. A. L., Chmurny, G. N., Krane, J.: J. Am. Chem. Soc. *95*, 4423 (1973).

133) Lambert, J. B., Mixan, C. E., Johnson, D H.: Tetrahedron Letters, *1972*, 4335.

134) Featherman, S. I., Quin, L. D.: J. Am. Chem. Soc. *95*, 1699 (1973).

135) Farnham, W. B.: J. Am. Chem. Soc. *94*, 6857 (1972).

136) Levin, R. H., Roberts, J. D., Kwart, H., Walls, F.: J. Am. Chem. Soc. *94*, 6856 (1972).

137) Vinter, J. G., Hoffmann, H. M. R.: J. Am. Chem. Soc. *95*, 3051 (1973).

138) Gittens, V. M., Eccleston, G., Wyn-Jones, E., Orville-Thomas, W. J.: Faraday Symp. Chem. Soc. *6*, 106 (1972).

139) Angiolini, L., Duke, R. P., Jones, R. A. Y., Katritzky, A. R.: J. Chem. Soc. Perkin II *1972*, 674.

140) Tavernier, D., Anteunis, M., Hosten, N.: Tetrahedron Letters *1973*, 75.

141) Yamamoto, O., Yamagisawa, M., Hayamishi, K., Kotowycz, G.: J. Mag. Res. *9*, 216 (1973).

142) Lambert, J. B., Mixan, C. E., Johnson, D. H.: J. Am. Chem. Soc. *95*, 4634 (1973).

Dynamics of Eight-Membered Rings in the Cyclooctane Class

Prof. Frank A. L. Anet

The University of California, Department of Chemistry, Los Angeles, California, USA

Contents

I. Introduction

This review deals with conformational interconversions in eight-membered cyclic compounds in the "cyclooctane class". It is convenient to classify ring systems according to the degree of endocyclic unsaturation that is present. In this context, *endocyclic* unsaturation can be a full double bond as in cyclooctene, a partial double bond as in enantholactam, a fused three-membered ring, as in cyclooctene epoxide or a triple bond as in cyclooctyne. *Exocyclic* unsaturated compounds are placed in the same class as the corresponding saturated compounds, except in cases where the double bond is appreciably delocalized into an endocyclic position (*e.g.* lactones and lactams). Thus cyclooctane and cyclooctanone are grouped together, in what is called, for convenience, the cyclooctane class, which also includes heterocyclic compounds such as oxacyclooctane (oxocane). The desirability of this classification in a conformational discussion arises from the strong geometric constraint of endocyclic double bonds, which force certain ring torsional angles to be close to 0°. In the compounds grouped together in the cyclooctane class, the preferred *ring* torsional angle is close to 60° and the torsional barriers are only a few kcal/mole. There are, of course, borderline compounds, which fortunately are not common (*e.g. cis* fused four- and eight-membered rings). Bicyclic compounds such as [3.3.0]bicyclooctane or [3.3.1]bicyclononane, although containing saturated eight-membered rings, are so geometrically constrained that they will also be excluded from the present review.

In order to discuss conformational interconversions, a knowledge of ground conformational states is required, and this topic will form part of the review. The dynamics of eight-membered rings have been studied almost entirely by nmr methods, the sole exception being some mechanical relaxation measurements which will be discussed at the end of Section VII.A. Only a brief qualitative introduction of the dynamic nmr method will be given. Conformational interconversions are often discussed in terms of ring inversion and pseudorotation, and definitions of these terms, as used in this chapter, are presented and discussed in Section III. No special distinction is made in this review between the terms conformation, conformer, or form.

II. Principles of Dynamic NMR

For details on the use of dynamic nmr to obtain kinetic parameters, the reader is referred to standard texts [1] and other reviews.[2-5] For the present purpose, it is useful to remember that the nmr spectrum of conformations which interconvert rapidly on the so-called nmr time scale will show averaged chemical shifts and coupling constants. When the interconversion is slow, each conformation will give rise to its own characteristic spectrum. In the intermediate region, the spectrum is broadened, and suitable analysis [4] can give rate constants (or life times) and free energies of activation ($\Delta G^{\ddagger}$) for the interconversion.

Measurements of rate constants at more than one temperature enable calculations to be made of the Arrhenius activation energy, and of the enthalpy $\Delta H^{\ddagger}$ and entropy $\Delta S^{\ddagger}$ of activation. In most cases, the accuracy of the nmr data is not sufficient for meaningful values of these three parameters to be obtained [5], and in most of the experimental work to be presented in this chapter, only $\Delta G^{\ddagger}$ will be given. However, strain energy calculations, with few exceptions, refer to ΔH at absolute zero, and not to ΔG. Since $\Delta G = \Delta H - T\Delta S$, and entropy effects appear to be only a minor perturbation in most cases, a comparison of $\Delta G^{\ddagger}$ with $(\Delta H^{\ddagger})_{0°K}$ can be justified, at least as a first approximation.

The nmr time scale, referred to earlier, corresponds to the inverse of the *difference* in resonance frequencies of interconverting systems. For protons, frequency differences are generally in the range of 1 to 100 Hz, while the differences for ^{19}F can be an order of magnitude larger. For $\Delta G^{\ddagger} = 5$ kcal/mole, the interconversion rate constant will have a value appropriate for the intermediate spectral region at temperatures in the range of roughly $-180°$ to $-150°C$, depending on the precise system being studied. Since high-resolution nmr spectra become very difficult to observe below these temperatures because of slow molecular tumbling, dynamic nmr studies cannot give information on free energy barriers much below 5 kcal/mole, even in the most favorable situation.

Only averaged spectra will be observed above about $-130°C$ when the energy barrier is 5 kcal/mole or less. For barriers of 10 kcal/mole, coalescence temperatures are typically in the range of $-40°$ to $-80°C$, while a barrier of 15 kcal/mole will give rise to a broadened spectrum near room temperature.

III. Definitions of Ring Inversion and Ring Pseudorotation

The term *ring inversion* has been widely employed in six-membered rings to describe the change from one chair to the alternate chair, a process which results in an exchange of axial and equatorial substituents.[6] It has also been used extensively for analogous processes in other ring systems. Lambert [7] has recently suggested that the word "inversion" be restricted to the atom (*e.g.* nitrogen) case and that the word *reversal* be used for rings. However, these words have such general meanings that they are nearly always qualified (*e.g.* ring reversal), and thus there seems to be little disadvantage to the use of the terms atom inversion and ring inversion even where both processes take place in the same molecule.

171

We offer two definitions of ring inversion, one strict, and one extended. For a process to be ring inversion, in the strict definition, all of the ring dihedral angles must change their *signs*, but not their absolute magnitudes. The definition given above is unambiguous, but it does not apply strictly to many situations, *e.g.* the axial methylcyclohexane — equatorial methylcyclohexane interconversion. To cover such systems we use an extended definition of ring inversion: ring inversion causes all the signs of the ring dihedral angles to change (except that dihedral angles which are close to 0° or 180° may or may not change signs) and the absolute magnitudes of these angles are either unaffected or only slightly changed. A further extension is possible and is useful in discussing nmr results: the dihedral angles can be averaged over one or more rapidly interconverting conformations.

The term *pseudorotation* was first applied to cyclopentane [8]; like inversion, it has an atomic analogue in 5-coordinate compounds (*e.g.* PF_5).[9] The name means "false rotation", and it is therefore appropriate for any conformational process which results in a conformation superposable on the original, and which differs from the original in being *apparently* rotated about one or more axes. Pseudorotation, in analogy with real molecular and internal rotations, can be free, as in cyclopentane, or more or less hindered, as in cycloheptane and higher cycloalkanes. In moderately to severely hindered pseudorotation, it is appropriate to consider distinct stable conformations which are pseudorotation partners, and these cases are often amenable to study by dynamic nmr methods. When the barrier to pseudorotation is very low, or in the limit when pseudorotation is free, it is not really justified to talk about separate stable conformations (*e.g.* the C_2 and C_s forms of cyclopentane), because strictly there is only one conformation, and the pseudorotation is simply a molecular vibration.

Ring inversion (when strictly defined) of achiral conformations is nothing more than a pseudorotation. Ring inversion in the cyclohexane chair, for example, leaves the molecule *apparently* rotated by 60° along the C_3 axis. Nevertheless, in order to conform with common usage, we will exclude ring inversion from the definition of ring pseudorotation.

The above discussion deals with a strict definition of pseudorotation. As with ring inversion, it is desirable to have an extended definition of pseudorotation. Before we can do this, we need to introduce a suitable abstraction, which we will call the *ring skeleton* of a conformation. The ring skeleton is simply the geometrical figure corresponding to the conformation, with ring bonds considered as straight lines, and ring nuclei considered as the vertices of the figure. In the extended definition, pseudorotation (either free or hindered) is a process which results in the *apparent* rotation of the ring skeleton of the conformation (with minor changes in the ring skeleton being ignored) or (to extend the definition slightly to cover chiral conformations) which converts the ring skeleton to its mirror image (again with minor changes in the ring skeleton being ignored). As before, it is useful to exclude from this definition those processes which can be called ring inversions (in the extended definition).

The extended definitions for ring inversion and (ring) pseudorotation will be used in the present chapter, unless stated to the contrary. These definitions are independent of mechanism, unlike Hendrickson's usage of these terms.[10-13]

IV. Structural Data

Cyclooctane itself has been studied by numerous physical methods, but the usual methods of structural determinations have proven singularly unrewarding. The molecule contains too many atoms for a reliable structural analysis by infrared and Raman spectroscopy [14]; all that can be safely said is that the molecule is not centrosymmetric. An X-ray diffraction study of crystalline cyclooctane (mp 14.4 °C) at 0 °C gave limited information because the crystals are disordered.[15] Electron diffraction measurements on cyclooctane vapor at 40 °C could not be reconciled with any single rigid conformation, but seemed to indicate a mixture of forms (or at least a very flexible conformation) which could not be defined further.[16] Nuclear magnetic resonance (nmr) studies at -130 °C exclude any conformation with a C_2 axis passing through carbon atoms, or a C_s plane passing only through bonds.[17] Interestingly, cyclooctane has been reported to have a small, but definite dipole moment in the gas phase.[18]

Table 1. X-ray structures of some eight-membered rings

Compound	Structure	Conformation and positions of substituents[1]	Dihedral angles for cyclooctane derivatives (in °)[2]	Ref.
trans-1,2-Cyclooctane carboxylic acid		Boat-chair (BC-4e,5e)	$\omega_1 = 62$ $\omega_2 = 47$ $\omega_3 = -106$ $\omega_4 = 71$ $\omega_5 = -70$ $\omega_6 = 101$ $\omega_7 = -43$ $\omega_8 = -63$	19)

Table 1 (continued)

Compound	Structure	Conformation and positions of substituents[1]	Dihedral angles for cyclooctane derivatives (in °)[2]	Ref.
cis-1,2- Cyclooctane carboxylic acid		Boat-chair (BC-2e',3e)	$\omega_1 = 60$ $\omega_2 = 48$ $\omega_3 = -105$ $\omega_4 = 67$ $\omega_5 = -66$ $\omega_6 = 96$ $\omega_7 = -38$ $\omega_8 = -69$	20)
trans-1,4- Dichloro cyclooctane		Boat-chair (BC-2e',5a)	$\omega_1 = 62$ $\omega_2 = 48$ $\omega_3 = -101$ $\omega_4 = 60$ $\omega_5 = -63$ $\omega_6 = 100$ $\omega_7 = -41$ $\omega_8 = -68$	21)
3,6-Spirooctylidene- 1,2,4,5-tetraoxacyclo- hexane		Boat-chair (BC-2e',2a')	$\omega_1 = 70$ $\omega_2 = 37$ $\omega_3 = -98$ $\omega_4 = 66$ $\omega_5 = -64$ $\omega_6 = 101$ $\omega_7 = -49$ $\omega_8 = -61$	22)
1-Aminocyclooctane carboxylic acid hydrobromide		Boat-chair (BC-2e',2a')	$\omega_1 = 67$ $\omega_2 = 41$ $\omega_3 = -98$ $\omega_4 = 67$ $\omega_5 = -65$ $\omega_6 = 98$ $\omega_7 = -46$ $\omega_8 = -63$	23)

Table 1 (continued)

Compound	Structure	Conformation and positions of substituents[1]	Dihedral angles for cyclooctane derivatives (in °)[2]	Ref.
1-Acetonyl-1-thionia-5-thia-cyclooctane perchlorate		Boat-chair (BC-3,7)		24)
5-Methyl-1-thia-5-azacyclooctane-1-oxide perchlorate		Boat-chair (BC-3,7)		25)
Azacyclooctane hydrobromide		(Boat-chair)[3] (or crown)	—	26)
Tetrathiacane		Boat-chair (BC-2,4,6,8)		27)

Table 1 (continued)

Compound	Structure	Conformation and positions of substituents[1]	Dihedral angles for cyclooctane derivatives (in °)[2]	Ref.
trans-syn-trans- 1,2,5,6- Tetrabromocyclooctane		Twist-chair-chair (TCC-1e,2e,5a,6a)	$\omega_1 = 60$ $\omega_2 = -79$ $\omega_3 = 107$ $\omega_4 = -72$ $\omega_5 = 42$ $\omega_6 = -77$ $\omega_7 = 116$ $\omega_8 = -81$	28)
Metaldehyde or all-*cis*-tetramer of acetaldehyde		Crown 1e,3e,5e,7e		29)
N—N'-Dimethyl-3,7-dithia-1,5-diazacyclo-octane		Crown		30)
1,3,5,7-Tetranitro-1,3,5,7-tetraaza-cyclooctane		Crown		31)

Table 1 (continued)

Compound	Structure	Conformation and positions of substituents[1]	Dihedral angles for cyclooctane derivatives (in °)[2]	Ref.
3,7-Dimethyl-1,5-dioxa-3,7-diazacyclo-octane-2,4,6,8-tetra-spirocyclopropane		Crown		32)
1,3,5,7-Tetrathia-2,4,6,8-tetraazacyclo-octane		Crown		33)
Octasulfur		Crown		34)

[1] Positions refer to the numbering of the carbon skeletons as shown in Fig. 1 and Fig. 2 (see also Table 3).

[2] The dihedral angles ω_1 and ω_2, etc. refer to positions 8,1,2,3; 1,2,3,4; etc. (i.e. they are the ring torsional angles about the 1,2 bond, the 2,3-bond, etc.).

[3] The X-ray data, found to fit a crown structure in early work, were later found to fit the boat-chair equally well.

In contrast to the parent compound, several cyclooctane derivatives and related compounds have had their structures determined by X-ray diffraction (Table 1). Most of these compounds have boat-chair conformations, but *trans-syn-trans*-1,2,5,6-tetrabromocyclooctane is a twist chair-chair, and crown conformations are found in octasulfur, the all-*cis* tetramer of acetaldehyde, and related compounds.

V. Strain Energy Calculations of Static Conformations

Semi-empirical strain energy calculations for cyclooctane have been carried out by four groups [10-13,35-37] (Table 2). The perspective drawings [38] in Fig. 1 were drawn by the computer program Ortep [39] with the parameters calculated by Hendrickson.[11] Table 2 gives dihedral angles, the sets of

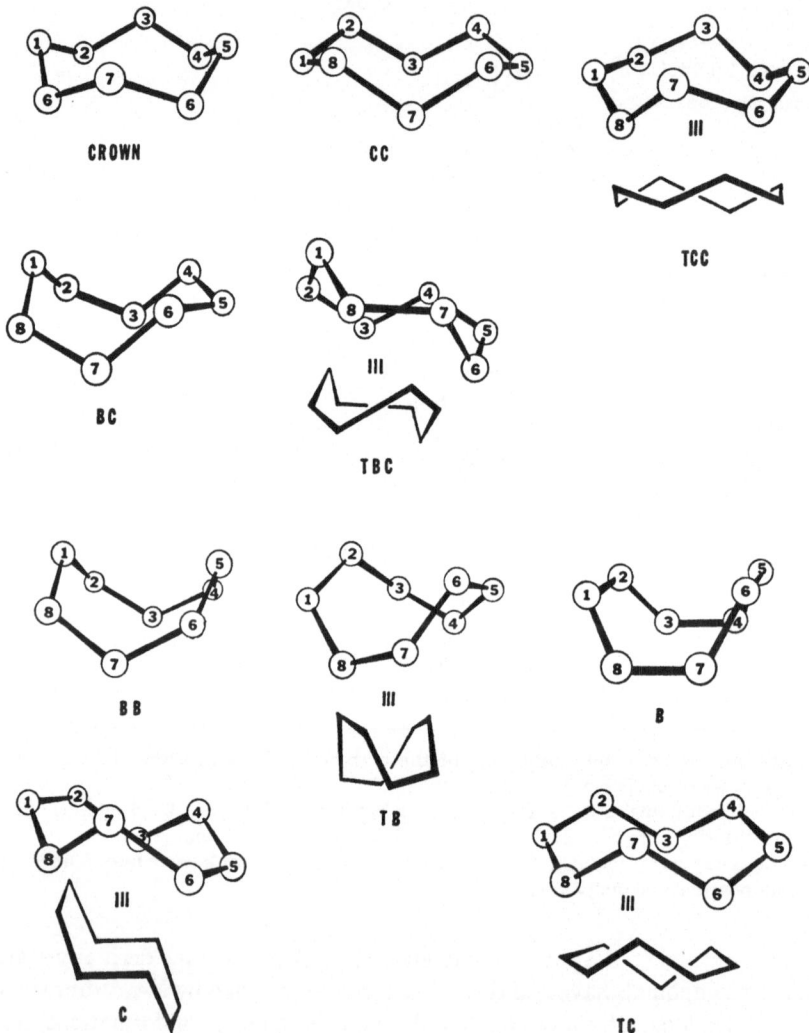

Fig. 1. Symmetrical conformations of cyclooctane [38], according to the calculations of Hendrickson.[11] Alternate views of some conformations are given

isochronous carbons[a], the symmetry groups, and the strain energies as calculated by the various workers. It should be noted that all the conformations in Table 2 have at least one symmetry element and that Wiberg [35] and Hendrickson [11] have presented data and arguments indicating that energy minima will occur in symmetrical conformations. The names given for the conformations are those proposed by Roberts in a recent paper.[40] Hendrickson's S_4 form is identical with the twist-boat of Roberts and this is the only difference between the nomenclature of the two investigators.[b]

A particular conformational name will be applied also to related conformations. For example, the boat-chair name will be applied to conformations of C_s symmetry (or to forms with ring skeletons having only slightly distorted C_s geometry), provided that the dihedral angles have the same *sign sequence* (and in general the same magnitude) around the ring as shown for the BC form in Table 2.

Three factors contribute to the strain energies in the conformations of cyclooctane: (a) eclipsing strains; (b) internal angle strain; and (c) non-bonded repulsions (and attractions). No conformation of cyclooctane is strain-free. The "ideal" or diamond-lattice form is the boat-boat, but this suffers from large non-bonded repulsions, which would be even larger were these repulsions not partly relieved by an increase in the internal angle to 118° or 119°. The crown form on the other hand has mainly eclipsing strain, which can be partly relieved by introducing distortions which lead to the chair-chair and the twist-chair-chair forms. The boat-chair and twist-boat-chair both have intermediate amounts of torsional and non-bonded strains. The chair, twist-chair, and boat forms are all highly strained and are mainly of interest as potential transition states for the interconversion of other conformations.

The internal angles in all the conformations given in Table 2 are considerably greater than the 111.5° found in cyclohexane. In the crown, chair-chair, twist-chair-chair, boat-chair, and twist-boat-chair, the angles are in the range of 115° to 117°. The experimentally determined internal and dihedral angles for several boat-chair cyclooctane derivatives (Table 1) are very similar to those calculated by Hendrickson and by Bixon and Lifson.

---

[a] Isochronous nuclei have the same nmr chemical shift by symmetry.

[b] According to Hendrickson's nomenclature the twist-boat ≡ boat-boat, and the twist-boat-boat ≡ boat. Although this usage is perfectly logical, it leaves the important S_4 form, which is intermediate between the boat-boat and the boat without a convenient name, and for this reason we prefer Roberts' nomenclature. Dunitz [26] has deplored the use of names such as "boat-chair" and has emphasized the importance of dihedral angles in defining conformations. Our view is that names are highly convenient in written and verbal discussions, and that the *naming* and the *definition* of a conformation are different things.

F. A. L. Anet

Table 2. Symmetrical conformations of cyclooctane

Conformation	Symmetry group	Dihedral angles[1] in °	Sets of isochronous carbons	H[2]	BL[3]	W[4]	A[5]
Crown	D_{4d}	$\omega_1 = \omega_3 = \omega_5 = \omega_7 = 87.5$ $\omega_2 = \omega_4 = \omega_6 = \omega_8 = -87.5$	One set	2.8	3.62	0.26	2.09
Chair-chair (CC)	C_{2v}	$\omega_1 = \omega_5 = -66.0$ $\omega_2 = \omega_6 = 105.2$ $\omega_3 = \omega_7 = -105.2$ $\omega_4 = \omega_8 = 66.0$	I: C_1, C_5 II: C_2, C_4, C_6, C_8 III: C_3, C_7	1.9	—	—	2.25
Twist-chair-chair (TCC)	D_2	$\omega_1 = \omega_5 = 56.2$ $\omega_2 = \omega_4 = \omega_6 = \omega_8 = -82.4$ $\omega_3 = \omega_7 = 114.6$	I: C_1, C_4, C_5, C_8 II: C_2, C_3, C_6, C_7	1.7	1.89	−0.25	2.20
Boat-chair (BC)	C_s	$\omega_1 = \omega_4 = 65.0$ $\omega_2 = 44.7$ $\omega_3 = -102.2$ $\omega_5 = \omega_8 = -65.0$ $\omega_6 = 102.2$ $\omega_7 = -44.7$	I: C_1 II: C_5 III: C_2, C_8 IV: C_3, C_7 V: C_4, C_6	0	0	0	0
Twist-boat chair (TBC)	C_2	$\omega_1 = \omega_5 = 51.9$ $\omega_2 = \omega_4 = 44.8$ $\omega_3 = -115.6$ $\omega_6 = \omega_8 = -93.2$ $\omega_7 = 88.0$	I: C_1, C_6 II: C_2, C_5 III: C_3, C_4 IV: C_7, C_8	2.0	—	—	—
Boat-boat (BB)	D_{2d}	$\omega_1 = \omega_2 = \omega_5 = \omega_6 = 52.5$ $\omega_3 = \omega_4 = \omega_7 = \omega_8 = -52.5$	I: C_1, C_3, C_5, C_7 II: C_2, C_4, C_6, C_8	1.4	—	4.44	—
Twist-boat (TB)	S_4	$\omega_1 = \omega_5 = -37.6$ $\omega_2 = \omega_6 = 64,9$ $\omega_3 = \omega_7 = 37.6$ $\omega_4 = \omega_8 = -64.9$	I: C_1, C_3, C_5, C_7 II: C_2, C_4, C_6, C_8	0.9	—	—	—

Table 2 (continued)

Conformation	Symmetry group	Dihedral angles[1] in °	Sets of isochronous carbons	Calculated relative strain energies in kcal/mole H[2] BL[3] W[4] A[5]			
Boat (B)	D_{2d}	$\omega_1 = \omega_3 = \omega_5 = \omega_7$ $= 0.0$ $\omega_2 = \omega_6 = \quad 73.5$ $\omega_4 = \omega_8 = -\ 73.5$	One set	10.3	—	8.96	—
Chair (C)	C_{2h}	$\omega_1 = \omega_3 = -\ 76.2$ $\omega_2 = \quad 119.9$ $\omega_4 = \omega_8 = \quad 0.0$ $\omega_5 = \omega_7 = \quad 76.2$ $\omega_6 = \quad -119,9$	I: C_1, C_2, C_5, C_6 II: C_3, C_4, C_7, C_8	8.3	—	6.09	—
Twist-chair (TC)	C_{2h}	$\omega_1 = \omega_8 = \quad 37.3$ $\omega_2 = \omega_7 = -109.3$ $\omega_3 = \omega_6 = \quad 109.3$ $\omega_4 = \omega_5 = -\ 37.3$	I: C_1, C_5 II: C_2, C_4, C_6, C_8 III: C_3, C_7	8.7	—	—	—

[1] See Fig. 1 for conformational drawings.
[2] From Ref.[11].
[3] From Ref.[35].
[4] From Ref.[36].
[5] From Ref.[37].

Table 3 gives Hendrickson's labelling scheme for the external bonds and substituents and also gives the excess strain energy caused by a methyl group at various positions in the cyclooctane conformation. Carbon atoms which are flanked by identical dihedral angles lie on C_2 axes and have identical *isoclinal* external bonds. If these dihedral angles are not identical, but still have the same sign and similar magnitudes, the two external bonds become slightly different and are labelled quasi-axial (*a'*) and quasi-equatorial (*e'*). All other carbons have quite different external bonds, which are labelled axial (*a*) and equatorial (*e*). The terms axial and equatorial as used here do not necessarily imply the same features as are found in the cyclohexane chair form. Nevertheless, a single axial methyl group always results in a substantial excess strain energy (> 1.0 kcal/mole), while a single equatorial methyl group gives rise to a low strain energy (0.3 to 0.8 kcal/mole). Isoclinal, *a'* and *e'* methyl groups also have low strain energies.

The labelling of hydrogens in the important boat-chair form of cyclooctane is shown in Fig. 2. Ignoring for the moment the distinction between

Table 3. Excess strain energy for methylcyclooctane

Conformation	Position of methyl group and stereochemical relationship[1])	Excess strain energy in kcal/mole[2])
Crown	1e	$\simeq 0.5$[3])
	1a	$\geqslant 1.5$[3])
Chair-chair (CC)	1a (α)	1.5
	1e (β)	0.4
	2e (α)	0.4
	2a (β)	3.1
	3a (α)	10.2
	3e (β)	0.4
Twist-chair-chair (TCC)	1e (α)	0.4
	1a (β)	1.5
	3a (α)	5.5
	3e (β)	0.5
Boat-chair (BC)	1a (α)	8.0
	1e (β)	0.5
	2e′ (α)	0.5
	2a′ (β)	0.5
	3e (α)	0.5
	3a (β)	7.6
	4e (α)	0.5
	4a (β)	5.1
	5a (α)	1.4
	5e (β)	0.6
Twist-boat-chair (TBC)	1e (α)	0.3
	1a (β)	7.2
	2e′ (α)	0.4
	2a′ (β)	0.5
	3a (α)	5.8
	3e (β)	0.5
	4e (α)	0.6
	4a (β)	3.0

Table 3 (continued)

Conformation	Position of methyl group and stereochemical relationship[1])	Excess strain energy in kcal/mole[2])
Boat-boat (BB)	1a	8.1
	1e	0.6
	2 isoclinal	0.6
Twist-boat (TB)	1a' (α)	0.6
	1e' (β)	0.5
	2a (α)	7.4
	2e (β)	0.5
Boat (B)	1a	5.1
	1e	0.8
Chair (C)	1a (α)	1.4
	1e (β)	0.7
	2e (α)	7.4
	2a (β)	0.4
Twist-chair (TC)	1 isoclinal	0.4
	2a (α)	4.6
	2e (β)	0.4
	3e (α)	0.3
	3a (β)	8.8

[1]) α and β are used here only to indicate *relative* stereochemistry within a conformation: a pair of α, or a pair of β hydrogens are always *configurationally cis*, while an α and β hydrogen are always *configurationally trans*.

[2]) From Ref. [12]).

[3]) Calculations for the crown were not reported by Hendrickson; the values given are rough estimates based on non-bonded distances in the crown.

a and *a'* or *e* and *e'*, we can see from this figure and from Table 3 that vicinal *trans* substituents are either diaxial or diequatorial (as in the cyclohexane chair), *except* when the substituents are on C-2 and C-3 (or the equivalent C-7 and C-8), in which case one substituent is axial and the other equatorial. This particular feature has very important consequences for pseudorotation in the boat-chair, and will be discussed further in the next section.

Allinger and coworkers have carried out strain-energy calculations on cyclooctanone.[41–43] Their latest results [43], which are given in Table 4,

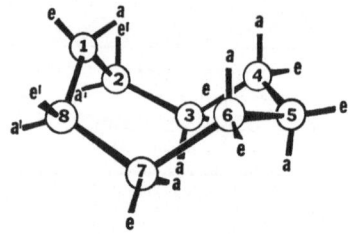

Fig. 2. Axial (*a*), equatorial (*e*), quasi-axial (*a'*) and quasi-equatorial (*e'*) positions in the boat-chair

show that the boat-chair with the carbonyl group at position 3 (*i.e.* BC-3) is the lowest energy conformation.

Strain energy calculations for other conformations in the cyclooctane class are unfortunately not available. In particular, there are no calculations for heterocyclic eight-membered rings. Finally, there is a need for more accurate and reliable calculations, which can give not only the equilibrium geometry and the strain energy, but also the vibrational frequencies. Only a very limited amount of work has been done along these lines.[44,45]

Table 4. Relative strain energies in cyclooctanone Conformations

Conformation	Strain energy (kcal/mole[1])
BC-1	2.9
BC-2	1.7
BC-3	0.0
BC-4	2.2
BC-5	3.9
TCC-2	1.5

[1] Relative to BC-3.

VI. Theoretical Interconversion Pathways

Hendrickson has investigated symmetrical interconversion pathways between cyclooctane conformations.[13] In some cases at least, these paths are probably not as low-energy as unsymmetrical paths, which unfortunately are more difficult to calculate. In the crown family, which includes the chair-chair and twist-chair-chair forms, the situation is quite clear. The twist-chair-chair is of lowest energy and conversion to the chair-chair form involves only the energy difference between the two forms. Thus the chair-chair is a transition state for the degenerate interconversion of twist-chair-

chair forms, and the activation energy is calculated to be only 0.2 kcal/mole. The crown itself is an energy maximum on the energy surface for distortion to the chair-chair and twist-chair-chair forms, and lies only 1.1 kcal/mole above the latter conformation. Therefore, all the conformations of the crown family will interconvert extremely rapidly, certainly far too rapidly for the interconversions to be studied by nmr line shape methods, even at −200 °C. For cyclooctane itself, this means that for nmr purposes the effective (i.e. time-average) symmetry is that of the crown, which contains only one kind of carbon, and only two kinds of hydrogens. Furthermore, vicinal *trans* hydrogens are either diaxial or diequatorial, just as in the cyclohexane chair. Thus, interconversion of the twist-chair-chair through the chair-chair (and/or crown) results only in partial averaging, because all the hydrogens are not isochronous, but form two distinct sets, one set including all the axial protons and the other set all the equatorial protons.

For all the protons in the twist-chair-chair to be isochronous requires interconversion with forms not in the crown family. Any rapid interconversion of the twist-chair-chair with either the boat or chair or boat-boat or twist-chair automatically assures that all the protons become isochronous. Interconversion of the twist-chair-chair with the boat-chair (and/or twist-boat-chair) alone does not necessarily lead to a single set of protons.

Interconversions in *the boat family*, which includes the boat, twist-boat, and boat-boat forms, have been extensively discussed by Hendrickson [13] and by Roberts and coworkers.[40] The twist-boat is calculated to be of lowest energy in this family [13], and it also has the lowest symmetry, having four different sets of isochronous protons. Interconversion with the boat-boat, which has only three different sets of isochronous protons, reduces the four sets to three, and the time-average symmetry becomes the same as that of the boat-boat form. The boat-boat form has 8 isoclinal protons on alternate carbons, and 4 axial and 4 equatorial protons situated in pairs, also on alternate carbons. This form is calculated to be the transition state for the interconversion of two twist-boats, and the activation energy is therefore only 0.5 kcal/mole (Table 2). Thus, for cyclooctane itself, twist-boat forms will interconvert extremely rapidly with one another, such that for nmr purposes the symmetry will appear the same as that of the boat-boat form, even at −200 °C.

The twist-boat can also interconvert with the boat form itself. The effect is to make all the protons isochronous if the twist-boat is also in rapid equilibrium with the boat-boat form, as is extremely likely from the previous discussion. The boat is probably the transition state in this pathway (TB → B → TB) and thus the activation energy calculated is 9.4 kcal/mole. This pathway has a sufficient barrier for study by nmr line shape methods if cyclooctane actually exists in these forms. At high temperatures (*e.g.* room temperature) all the protons should be isochronous and should give rise to a single line. At sufficiently low temperatures (*e.g.* $< -130\,°C$), the spectrum is expected to be very complex because of coupling between three different kinds of protons. If a deuterated cyclooctane with just two geminal protons is examined by proton nmr, with the deuterons decoupled, a particularly clear spectrum should be observed. In one half of the molecules (neglecting any small isotope effect) the protons will be isoclinal and will therefore give rise to an A_2 spectrum, *i.e.* a single line of total intensity 2; in the remaining molecules the protons will be on carbons which bear axial and equatorial bonds, and will therefore give rise to an AB spectrum, *i.e.* four lines of total intensity 2. If the AB spectrum is accidentally degenerate, it will appear as a single line of intensity 2. It is impossible, however, for the spectrum to appear as a *single AB quartet* (experimental data on this point will be discussed later, but it may be mentioned here that a single AB quartet is in fact observed for the compound under discussion). If cyclooctane were to exist in a crown family form, this deuterated cyclooctane would be expected to give a single AB quartet at low temperatures. Thus, the observation of a single AB quartet in this case rigorously eliminates the boat-boat (and with very high probability the twist-boat also). The results, although compatible with a crown family form, do not require such a form. As we shall see later, a boat-chair conformation also fits the experimental data of a single AB quartet.

Interconversions which involve the *boat-chair* and *twist-boat-chair* (Table 5) are important because most simple cyclooctane derivatives have boat-chair conformations. Table 5 shows the effect of a rapid equilibrium between the boat-chair and other conformations, either singly or in various combinations. Some of these interconversions are unlikely to be important, but are given for completeness. For example, the interconversion of the boat-chair with the chair-chair form *alone* is improbable, since the activation energy for this process is undoubtedly very much larger than the barrier for the chair-chair to twist-chair-chair interconversion, which has already been discussed. A rocking motion of the boat-chair C-1 methylene group can lead to the chair-chair, a plane of symmetry being maintained during the change. Such a symmetrical path is calculated [13] to have a very high energy (*ca.* 20 kcal/mole) but lower energy paths exist to the crown family. The path BC → TBC → TCC has a calculated activation energy of 11.4 kcal/mole, with the "transition state" being between the twist-boat-chair and the twist-chair-chair.[13] This "transition state" has the same symmetry as the regular twist-chair-chair, but with $\omega_{12} = \omega_{56} = 0°$, and can also lead to the boat conformation *via* Roberts' [40] "parallel boat", a conformation not shown in Table 2.

One of the most important interconversion paths shown in Table 5 is the pseudorotation of the boat-chair *via* the twist-boat-chair as an intermediate. This pseudorotation was first recognized by Anet and St. Jacques [17,46], and leads to the same averaging as the boat-chair to twist-chair-chair inter-conversion. Unfortunately, strain energy calculations are not available for the boat-chair pseudorotation, but qualitative considerations [13,17,46] indicate that the barrier to pseudorotation should be quite low, perhaps a few kcal/mole.

Dreiding-Fieser models can be used to demonstrate the interconversion of the boat-chair to the twist-boat-chair, but some practice is required to hold the model correctly. The motions involved in the pseudorotation are shown in Fig. 3, which shows one half of a full pseudorotation, namely the conversion of the boat-chair to the twist-boat-chair. The second half of the pseudorotation is exactly the reverse of the first half, except that it is carried out after rotation of the twist-boat-chair by 180° about its C_2 axis.

The twist-boat-chair is chiral and hence there are actually two mirror-image or enantiomeric conformations. Correspondingly, there are two mirror-image pseudorotation paths to the twist-boat-chairs from the boat-chair, which is achiral. These paths are conveniently labelled by the mirror-image letters b and d. If we arbitrarily assign the letter b to the process shown in Fig. 3, and if we use the proton in the 1-equatorial position as a label, we can write down the effect of pseudorotation as shown in Table 6.

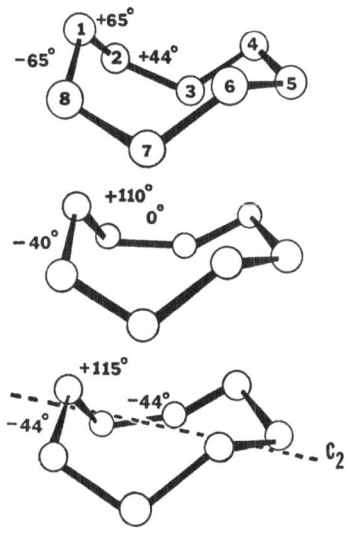

Fig. 3. Interconversion of the boat-chair (top) to the twist-boat-chair conformation (bottom)

Table 5. Nmr effects in the boat-chair

Conformations in rapid equilibrium with the boat-chair (BC)	Sets of isochronous nuclei		Time-average symmetry
	^{13}C	^{1}H	
Twist-boat-chair (TBC) [1] and/or crown and/or twist-chair-chair (TCC)	One set	A: $1e, 2e', 3a, 4a,$ $5a, 6a, 7a, 8e'$ B: $1a, 2a', 3e, 4e,$ $5e, 6e, 7e, 8a'$	D_{4d}
Chair-chair (CC)	I: $1,5$ II: $2,4,6,8$ III: $3,7$	A: $1e, 5a$ B: $1a, 5e$ C: $2e', 4a, 6a, 8e'$ D: $2a', 4e, 6e, 8a'$ E: $3e, 7e$ F: $3a, 7a$	C_{2v}
Twist-chair (TC)	I: $1,5$ II: $2,4,6,8$ III: $3,7$	A: $1e, 5e$ B: $1a, 5a$ C: $2e', 4e, 6e, 8e'$ D: $2a', 4a, 6a, 8a'$ E: $3e, 3a, 7e, 7a$	C_{2h}

Conformation				Symmetry
Boat-boat (BB) and/or twist-boat (TB) [2]	I: 1,3,5,7 II: 2,4,6,8	A: 1e, 3e, 5a, 7e B: 1a, 3a, 5e, 7a C: 2e', 2a', 4e, 4a, 6e, 6a, 8e', 8a'		D_{2d}
Boat (B) or chair (C) { twist-boat-chair or crown or twist-chair-chair } and { boat-boat or twist-boat or twist-chair } [3]	One set	One set		D_{8h}
Chair-chair (CC) and twist-chair (TC)	I: 1,5 II: 2,4,6,8 III: 3,7	A: 1e, 1a, 5e, 5a B: 2a', 2e', 4a, 4e, 6e, 6a, 8e', 8a' C: 3e, 3a, 7e, 7a		D_{2h}

1) Additional averaging through CC has no further effect.

2) Additional averaging through CC and/or TC has no further effect.

3) These equilibria lead to complete averaging; combination of any one of these equilibria with any other given in this table, will of course also lead to complete averaging.

Table 6. Boat-chair pseudorotation (*via* TBC) in a monosubstituted cyclooctane

Pseudorotation step(s)	Effect on 1e substituent
b	$1e \longrightarrow 2e'$
d	$1e \longrightarrow 8e'$
bb	$1e \longrightarrow 2e' \longrightarrow 1e$
dd	$1e \longrightarrow 8e' \longrightarrow 1e$
bd	$1e \longrightarrow 2e' \longrightarrow 7a$
db	$1e \longrightarrow 8e' \longrightarrow 3a$

It can be seen that two pseudorotation steps of the same chirality have no net effect. A sequence of steps with alternating chiralities returns the conformation to its original state after eight steps as follows:

$$1e \xrightarrow{b} 2e' \xrightarrow{d} 7a \xrightarrow{b} 4a \xrightarrow{d} 5a \xrightarrow{b} 6a \xrightarrow{d} 3a \xrightarrow{b} 8e' \underset{d}{\longleftarrow}$$

This sequence of steps is a $(bd)_4$-pseudorotation itinerary. The mirror-image itinerary gives rise to the same sequence but with all the arrows reversed in direction. Although the 1-equatorial proton passes through various axial and equatorial sites, only one half of the boat-chair positions (*i.e.* set A in Table 5) are visited during a cycle. If we follow the 1-axial proton, we find that it visits the other half of the available positions (*i.e.* set B in Table 5) as follows:

$$1a \xrightarrow{b} 2a' \xrightarrow{d} 7e \xrightarrow{b} 4e \xrightarrow{d} 5e \xrightarrow{b} 6e \xrightarrow{d} 3e \xrightarrow{b} 8a' \underset{d}{\longleftarrow}$$

Replacement of a CH_2 group in cyclooctane by a CHX or CX_2 group, where X is achiral, or by a heteroatom such as O or S, leads to a *chemically* achiral molecule. If the conformational preference of the group or heteroatom is for the 1 or 5 positions in the boat-chair, the molecule is also conformationally achiral. However, if the preference is for any other positions, two enantiomeric conformers will result. In the latter case, dynamic nmr can be used to study the sequence of pseudorotations which leads to conformational racemization.

Experimental work, described in Section VII, shows that the barriers to pseudorotation are highly dependent on the nature of the CX_2 group, but unfortunately, no strain energy calculations are available on this point.

Examination of models, however, shows that a bulky CX_2 group, which can be accommodated at position 2 in the boat-chair, cannot be pseudorotated to the mirror-image position 8 without passing through very strained boat-chairs, *e.g.* the BC-1 conformation. Furthermore, the $\omega_{23} = 0$ conformation (Fig. 3) resulting from BC-1 is much more strained than is the corresponding form of cyclooctane itself.

At this point it becomes necessary to emphasize that *the number and letters (e.g. 1 or 1e)* used in the present discussion are part of a conformational labelling scheme, which must not be confused with any chemical labelling that may be required. In a monosubstituted cyclooctane, no chemical labelling is needed if the focus is on the substituent, just as in the description of pseudorotation presented above. If reference is made to a particular ring proton or carbon in a monosubstituted cyclooctane, however, a double labelling scheme becomes essential, and such a scheme is needed in general. In this chapter the capital letters A, B, C, ... will always refer to chemical labelling. When numbers are used, the context will usually indicate whether a chemical or conformational label is being used. Numbers with added letters (*e.g.* 1a, 1e) are always conformational labels.

Table 5 shows that rapid interconversion of the boat-chair with either the boat or chair is sufficient to average all the protons in cyclooctane to one set. The chair is an energy maximum for the process BC → TBC → C → TBC → BC, and the activation energy from Table 2 is 8.3 kcal/mole. The boat path appears to be higher in energy than the chair path.[13]

With cyclooctane and other eight-membered rings which have low barriers to pseudorotation, a complete averaging of the protons to one set can also be achieved by interconversion of the boat-chair with the twist-chair, boat-boat or twist-boat conformations. When pseudorotation is fast, the process which leads to complete averaging can be called ring inversion. Ring inversion through the twist-chair is calculated to have a higher activation energy (11.6 kcal/mole) than inversion through the chair. The boat-boat path has a very high activation energy (20 kcal/mole) if a plane of symmetry is maintained. However, an unsymmetrical path, BC → TBC → TB → BB → TB → TBC → BC, appears to be of much lower energy, but strain energy calculations are not available. From the experimental data on cyclooctane (Section VII. A) it is known that the process which leads to complete averaging must have an activation energy of about 8 kcal/mole, and therefore the boat-chair to boat-boat interconversion cannot have an activation energy lower than 8 kcal/mole (this assumes that cyclooctane exists in the boat-chair, as is highly probable). Thus all of the above paths appear to be reasonable mechanisms of ring inversion, at least in cyclooctane or simple derivatives. The presence of substituents can change the barriers for the various paths, so that all ring inversions may not necessarily have the same mechanism.

Recent experimental data on heterocyclic eight-membered rings (Section VII.C) show that the barrier to interconversion of the boat-chair with a crown family conformation is higher than the barrier for ring inversion in the boat-chair. The data also show that a crown family form does not undergo ring inversion without also interconverting with the boat-chair. The strain energy calculations [13] do not reveal any path for ring inversion in the crown family of lower energy than the interconversion of this family with the boat-chair, in agreement with the experimental data.

It appears clear from the strain energy calculations and the experimental evidence (Section VII) that, at least for cyclooctane itself, the conformations of low energies can be grouped into three families which are separated by relatively high barriers (8 to 11 kcal/mole). The members of a given family, however, are separated from one another by much smaller barriers (0 to perhaps 4 kcal/mole). Table 7 contains a summary of these conclusions.

Table 7. Families of rapidly interconverting low-energy cyclooctane conformations

Name of family	Members	Sets of isochronous nuclei	
		carbon	hydrogen
Crown	Crown, chair-chair, twist-chair-chair	1	2
Boat-chair	Boat-chair, twist-boat-chair	1	2
Boat-boat	Boat-boat, twist-boat	1	3 [1]

[1] In the ratio of 2:1:1.

Certain 1,3-dioxocanes containing *gem*-dimethyl groups seem to exist in boat-chairs which have lower barriers for interconversion to the boat-boat than for interconversion to the twist-boat-chair, *i. e.* just the reverse of the normal situation. A discussion of these rather special systems will be deferred to Section VII.C, where the experimental data will be presented.

VII. Experimental Data and Discussion

A. Cyclooctane and Simple Derivatives

Proton nmr spectra of various deuterated cyclooctanes, and of cyclooctane itself, show a single chemical shift at room temperature, and two chemical shifts below about $-120\,°C$. The free energy of activation, $\Delta G^{\neq}$, for this process, *i.e.* for ring inversion, is 8.1 kcal/mole at $-112\,°C.$[47] The values given [47] for $\Delta H^{\neq}$ (7.4 kcal/mole) and $\Delta S^{\pm}$ (-4 eu) may have somewhat higher errors than those quoted (± 0.3 kcal/mole and ± 2 eu respectively) because of systematic errors, as mentioned in Section II.

The spectra of the deuterated cyclooctanes show no further changes down to $-175\,°C$, apart from some line broadening ascribable to slow molecular tumbling at low temperatures. The spectra at low temperatures (say $-135\,°C$) of the specifically deuterated cyclooctanes I, II and III (Fig. 4) exclude any systems which have the same time-average symmetry as the boat or boat-boat.[17] For example, compound III, which has a single CH_2 group gives rise to a single AB quartet and, as discussed in Section VI, this cannot be reconciled with a boat-boat or a twist-boat. The spectra are compatible with any conformation in the crown or boat-chair families, with the assumption that the crown to twist-chair-chair to chair-chair interconversion and the boat-chair to twist-boat-chair interconversion are very rapid even at low temperatures. This assumption is strongly supported by the strain energy calculations already presented. The crown and boat-chair family conformations require that the spectrum of I, which has two *cis* vicinal protons, be an AB quartet, as is indeed observed (Fig. 4). The spectrum of II, which has two *trans* vicinal protons should consist of two lines (*i.e.* two A_2 spectra), again as observed.

Since strain energy calculations and experimental data strongly indicate that the barrier for interconversion from the crown family to the boat-chair family is of the order of 11 kcal/mole, it is not possible for cyclooctane to exist as a mixture of these two families, unless one family is present in such a small amount that its spectrum is lost in the noise. Thus, one family must be present to more than 95% in cyclooctane at $-130\,°C.$[c]

The nmr spectra of various rather *simple* cyclooctane derivatives discussed below are only consistent with the boat-chair conformation. Furthermore the structures in the crystalline state are boat-chairs (Table 1), with one exception which can be rationalized (see below). The conclusion that cyclooctane exists in solution as the boat-chair therefore appears inescapable.

One difficulty with the conformational picture presented above is that it does not immediately provide an explanation for certain pulse nmr measurements on cyclooctane carried out by Meiboom some ten years ago.[48] A plot of $\log\,(1/T_2 - 1/T_1)$ (T_1 and T_2 are the spin-lattice and transverse relaxation times respectively) against the reciprocal of the absolute temperature does not give a straight line in the temperature range $-50°$ to $-105°C$, and indicates the presence of two processes. Ring inversion in the boat-chair can be one of these processes, but the second process cannot be the boat-chair pseudorotation, which is far too rapid in this temperature range to lead to any observable effect. The following explanation [49] affords a pos-

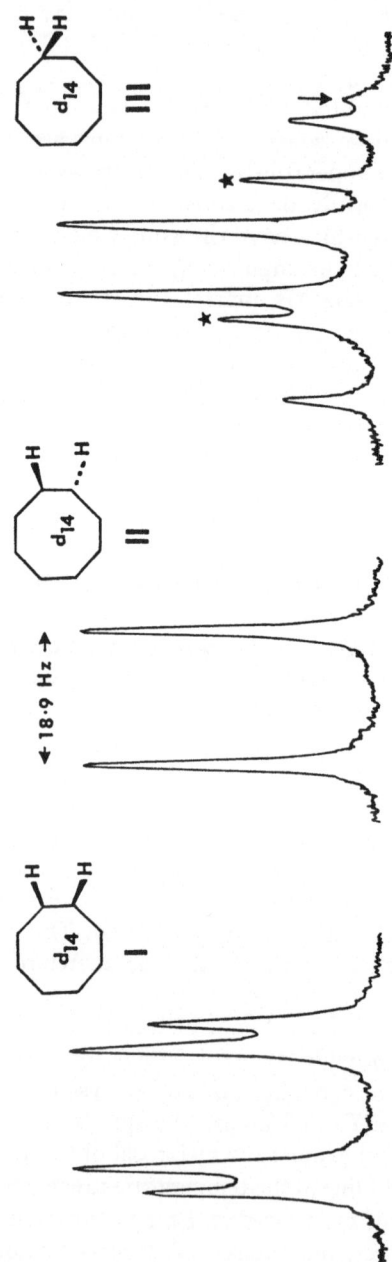

Fig. 4. 60 MHz Proton nmr spectra of three deuterated cyclooctanes at −140 °C.[17] The lines marked with stars arise from isotopic impurities having the partial structure —CD_2—CHD—CD_2—. The line marked with an arrow is a ^{13}C satellite of internal tetramethylsilane

sible way out of this dilemma. The second process could be a boat-chair to crown family (e.g. TCC) interconversion, in which the population of the twist-chair-chair is quite low (<5%), as is required from the previously described nmr evidence. The situation would then be like that of methyl-cyclohexane [50], where the interconversion of the axial form and the equatorial form is only observable under special conditions, because of the very unequal populations.[c]

Methylcyclooctane and t-butylcyclooctane both give strongly temperature-dependent nmr spectra in a range close to that observed for cyclooctane itself.[46,51] Such a behaviour is not expected if these compounds have crown-family conformations, because the alkyl group will take up equatorial positions almost completely, and as discussed in Section VI, all equatorial positions in this family are expected to interconvert extremely rapidly even at low temperatures, leading to (virtually) temperature-independent nmr spectra, as in monoalkylcyclohexanes. In the boat-chair, by contrast, pseudorotation via the twist-boat-chair gives rise to two sets of positions, previously referred to as sets A and B (Table 3). Both sets have unhindered equatorial positions (set B also has relatively unhindered quasi-axial positions), and interconversion between set A and set B is the ring inversion process just discussed in cyclooctane itself. In an alkyl-cyclooctane, ring inversion by any of the mechanisms mentioned in Section VI merely interconverts conformations with equatorial alkyl groups in set A with conformations with equatorial groups in set B. The alkyl group, it should be noted, remains unhindered during the entire ring inversion process, and thus the activation energy should be very similar to that in cyclooctane itself.

A crown-family conformation also cannot explain the nmr spectrum of the acetonide of trans-1,2-cyclooctanediol, a compound which has a five-membered ring fused to the cyclooctane ring.[46] A temperature-independent spectrum would be predicted, because a trans-fused five-membered ring can only be located at equatorial positions in crown-family conformations, and ring inversion is therefore prohibited. The nmr spectrum of this compound is actually strongly temperature-dependent at about −70 °C, thus excluding any conformation in the crown family, at least as the sole conformation.

In the boat-chair, both set A and set B have positions where a trans-fused five-membered ring can be located, and thus ring inversion is possible, and a temperature-dependent spectrum is allowed, in agreement with experiment. The ring inversion barrier is somewhat higher than in cyclooctane, but this could be due to the restraint caused by the five-membered ring. An explanation based on a mixture of twist-chair-chair and boat-chair conformations is also possible, but appears less likely.

---

[c] See section added in proof at end of text.

Symmetrical 1,1-disubstituted cyclooctanes provide very valuable conformational information. With fluorine as a substituent, in particular, it is possible to examine the ^{19}F resonance spectrum and to take advantage of the large fluorine chemical shifts.[40] Furthermore, although fluorine is appreciably larger than hydrogen, it is small enough to fit in not too hindered axial positions. Therefore, cyclooctane itself and 1,1-difluorocyclooctane (IV) should exist in the same conformation. The ^{19}F spectrum of IV, taken with protons decoupled, changes from a single line at room temperature to an AB quartet ($\delta_{AB} = 3.9$ ppm) at $-120\,°C$, and to two AB quartets at $-175\,°C$ ($\delta_{AB} = 14.3$ and 16.7 ppm, with intensities in the ratio of $2:1$ respectively).[40] The presence of two processes with free energies of activation of 7.5 and 4.9 kcal/mole is not compatible with crown-family conformations, which should show only a single process with an appreciable activation energy. An examination of the twist-chair-chair pseudorotation indicates that a CF_2 group should not increase the barrier by any significant amount over the value (0.2 kcal/mole) calculated for cyclooctane. The crown-family conformations also cannot explain the small chemical shift difference between the fluorine nuclei at $-120\,°C$.

Roberts [52] originally considered a twist-boat conformation for IV, but the mounting evidence for a boat-chair conformation for cyclooctane and various derivatives, led Roberts and coworkers to suggest boat-chair conformations for IV also.[40] Futhermore, the original explanation requires that pseudorotation of the twist-boat *via* the boat be of lower energy than the pseudorotation *via* the boat-boat and this is not supported by recent strain energy calculations.

The ^{19}F nmr spectra of IV can be nicely explained on the basis of the boat-chair conformation and the following reasonable assumptions:

a) The CF_2 group resides only at relatively unhindered boat-chair sites, *i.e.* only forms IV-BC-2, IV-BC-5, and IV-BC-8 are appreciably populated.

b) The free energy barrier to pseudorotation *via* the twist-boat-chair is 4.9 kcal/mole, and is thus probably somewhat higher than in cyclooctane itself, for reasons given in Section VI, and discussed in greater details below.

Forms IV-BC-2 and IV-BC-8 are enantiomeric and must have the same populations. Since pseudorotation of IV-BC-2 into IV-BC-8 does not cause an exchange of the quasi-axial with the quasi-equatorial fluorine, it is an invisible process by ^{19}F nmr, although this process can, in principle, be studied by ^{1}H or ^{13}C nmr. Pseudorotation of IV-BC-2 (or IV-BC-8) to

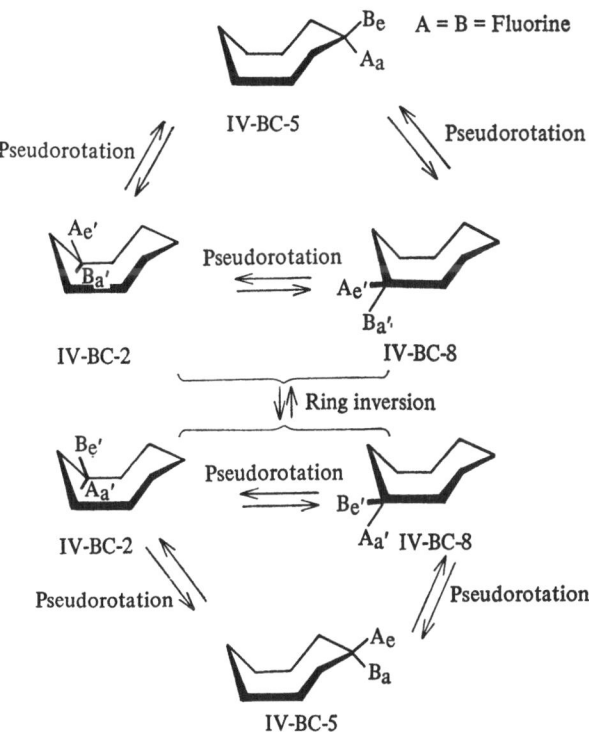

IV-BC-5 involves three pseudorotation steps and causes the following exchange of fluorines (shown only for the BC-2):

$$Ae' \rightleftarrows Aa$$
$$Ba' \rightleftarrows Be$$

The exchange of quasi-equatorial with axial fluorines provides an explanation for the small chemical shift difference at intermediate temperatures, *i.e.* when pseudorotation is fast.

Roberts and coworkers were able to simulate the experimental spectra over the temperature range $-160°$ to $-170°C$, using a computer program suitable for the exchange of two different AB systems. In this temperature range line widths in the absence of exchange are changing rapidly, so that relatively large errors could easily occur in calculations of kinetic parameters *other* than $\Delta G^\ddagger$, as pointed out by these authors. Thus no great significance can be attached to the calculated values of $\Delta S^\ddagger$ and $\Delta H^\ddagger$.

The process with a $\Delta G^\ddagger$ of 7.5 kcal/mole, results in the exchange of geminal fluorines and is a ring inversion of the kind already discussed. The intensity of 2:1 for the two AB systems in IV is not very well explained by invoking only three boat-chair forms, as Roberts and coworkers have done. The BC-2 and BC-8 forms are equal in energy, and have no non-bonded fluorine-hydrogen repulsions. The BC-5 should have a little fluorine-hydrogen nonbonded strain (see Table 3) and should therefore contribute *less* than 33% to the total. A possible explanation is that there is some population of the BC-4 and BC-6 conformations. These forms should be only somewhat less stable than the BC-5 conformation, and should interconvert extremely rapidly to the BC-5 by a single-step pseudorotation (see discussion of 1,1,2,2-tetrafluorocyclooctane conformations below). Thus the sum of the BC-4 and BC-5 and BC-6 conformations might well amount to one third of the total.

The ^{19}F nmr spectra of 1,1,4,4-tetrafluorocyclooctane (V) at low temperatures provides strong support for the interpretation of the spectrum of the 1,1-difluoro compound.[40] The spectrum of V changes upon lowering the temperature to a very strongly coupled AB system, which has not been investigated in detail. A further lowering of the temperature gives a second spectral change, as with IV, and at −160 °C, two nearly superposed AB quartets having the same intensity are observed. The $\Delta G^{\neq}$ for this process is 6.1 kcal/mole at −130 °C. From the previous discussion, V-BC-2,5 and V-BC-5,8, are the only unstrained boat-chairs (note that 2,5, and 8 are conformational labels). These forms explain the presence in V of two AB systems with equal intensities at low temperatures.

A three-step pseudorotation itinerary (*dbd*) transforms V-BC-2,5 into V-BC-5,8 and results in an exchange of fluorines very similar to the exchange in 1,1-difluorocyclooctane. Models show that at one point in the itinerary,

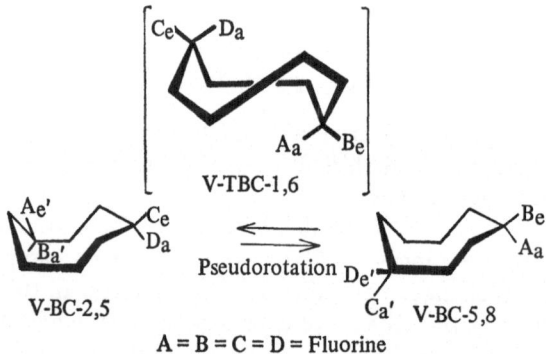

V-TBC-1,6

Pseudorotation

V-BC-2,5   V-BC-5,8

A = B = C = D = Fluorine

the system exists in the TBC-1,6 conformation, where *both* CF_2 have bad non-bonded repulsions. The free energy of activation for pseudorotation in V is, in fact, 1.2 kcal/mole greater than the corresponding value in 1,1-difluorocyclooctane. These results allow a rough calculation of the $\Delta G^{\neq}$

for pseudorotation in cyclooctane itself, which can be estimated to be $(\Delta G^{\pm})_{IV}$ −1.2, *i. e.* 3.7 kcal/mole.

Another compound investigated by Roberts and coworkers is 1,1,2,2-tetrafluorocyclooctane (VI), an isomer of V. The ^{19}F spectrum of VI changes upon lowering the temperature from a single line at room temperature to an AB quartet ($\delta_{AB} = 8.85$ ppm) plus a rather broad singlet almost centered on the quartet at −65 °C. The AB system and the single line have approximately the same intensities. Below −80 °C the single line broadens and below −110 °C becomes a second AB system almost superposed on the original AB quartet, which is unchanged.

Roberts and coworkers [40] interpreted the spectra of VI in terms of twist-boats (without counting enantiomers, two different twist-boats are possible with the substitution pattern present in VI). Each twist-boat (*e.g.* VI-TB) has four different sites when all pseudodoratations are slow, and under these conditions each form should give an ^{19}F spectrum consisting of an ABCD system (approximately two separate AB quartets). The two different TB forms were assumed to have approximately equal populations and to have coincidences in chemical shifts so that the expected four AB quartets would be virtually coincident. With the appropriate chemical shift assignment, it can then be shown [40] that pseudorotation of the two different twist-boats *via* a boat-boat results approximately in an AB quartet and a single central line. Further pseudorotation through the boat form, *i.e.* ring inversion, results in a single chemical shift for all fluorines in VI.

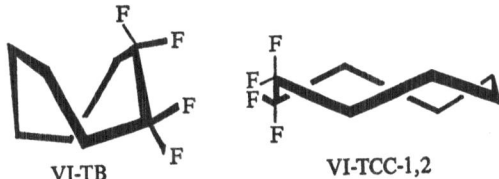

VI-TB   VI-TCC-1,2

The interpretation given above by Roberts and coworkers requires several coincidences, and is not free from objections. The difficulties with the twist-boat are as follows:

a) despite the rather low strain energy calculated for this conformation, no cyclooctane derivative or related compound is known to exist unequivocally in that conformation;

b) the strain energy for placing a CF_2 group at the 2 position of the twist-boat must be quite considerable (see Table 3 for methyl strain energies) and since the twist-boat is already not the most stable cyclooctane conformation, this extra strain does not help;

c) a rational alternative explanation in terms of the twist-chair-chair and the boat-chair can be formulated.

The boat-chair is clearly not a very good conformation for VI, because it is impossible to place the four fluorines in unhindered positions. In any case, as pointed out by Roberts and coworkers, a boat-chair conformation (alone) cannot explain the ^{19}F nmr data. Roberts and coworkers have quite correctly emphasized the rather large dipolar energy which results from eclipsing two CF_2 groups. The best conformation of VI should therefore be obtainable by minimizing (a) the fluorine non-bonded repulsions; (b) the dipolar repulsions; and (c) the inherent strain energy of the chosen cyclooctane conformation. The only conformations which fit the above restraints are the following: BC-4,5, TCC-1,2, CC-1,2 and TBC-7,8.

The BC-4,5 form has only one bad feature, namely, the strain of the $4a$ fluorine. A $4a$ methyl group on the boat-chair has a calculated strain energy of 5.1 kcal/mole (Table 3). The strain energy of a $4a$ fluorine must be much less than that of a methyl group, and a value in the range of 1 to 2 kcal/mole would seem reasonable.[53] The TCC-1,2 also has only one bad feature, namely, the higher strain energy of the TCC over the BC, and this amounts to 1.7 kcal/mole (see Table 3). The CC-1,2 should be rather similar in energy to the TCC-1,2; in any case these crown-family forms should interconvert very rapidly so as to give the effective symmetry of the TCC-1,2. It can now be seen that the BC-4,5 and the TCC-1,2 should be of about the same energies and might occur in about equal amounts.

The BC-4,5 can undergo pseudorotation to give back the same form, except that the CF_2 groups have interchanged positions. However, axial and equatorial fluorines do not exchange, nor do the CF_2 groups become eclipsed during the single-step pseudorotation. Such a pseudorotation should therefore have a $\Delta G^{\ddagger}$ scarcely higher than that ($\simeq 4$ kcal/mole) in cyclooctane itself, and thus would be rapid at all temperatures at which the spectrum of VI was investigated. Ring inversion of the BC-4,5 conformation, which can take place by pseudorotation *via* a chair form, followed (or preceded) by several steps of comparatively low-energy pseudorotation *via* the twist-boat-chair, gives rise to the BC-5,6 conformation. The BC-5,6 is the mirror-image of the BC-4,5, and will therefore show the same kind of rapid pseudorotation described earlier for the BC-4,5. Therefore, the BC-4,5 and BC-5,6 conformers, will give rise to a single AB system at low temperatures, and to a single line at high temperatures. These changes can occur without any interconversion to the TCC-1,2.

The TCC-1,2 is axially symmetric and will therefore give an AB system when ring inversion is slow, and a single line when ring inversion is fast. Furthermore, the barrier to ring inversion is expected to be higher in the twist-chair-chair than in the boat-chair conformation, both from strain energy calculations and by analogy with other compounds known to contain crown-family conformations.

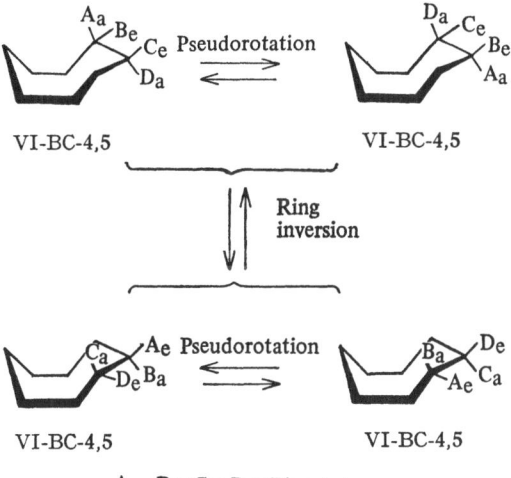

VI-BC-4,5   VI-BC-4,5

Ring inversion

VI-BC-4,5   VI-BC-4,5

A = B = C = D = Fluorine

Thus, the AB quartet in VI at −65 °C can be assigned to the TCC-1,2, which, at that temperature, is undergoing slow ring inversion; the single broad line in VI at −65 °C can be assigned to the BC-4,5 which is inverting moderately rapidly to the BC-5,6, with both boat-chairs undergoing very rapid single-step pseudorotations. At −109 °C, ring inversion has become slow in the boat-chair, but pseudorotation is still fast.

The two alternate explanations for the low-temperature nmr spectra of VI are potentially distinguishable. Even the ^{19}F spectra are not quite the same: the twist-boats should give two ABCD systems (with suitable co-incidences in chemical shifts), while the 1:1 mixture of the boat-chair and twist-chair-chair should give two different AA'BB' systems. Unfortunately, vicinal fluorine-fluorine coupling constants are extremely small, so that two apparent AB systems are expected from both explanations.

Roberts and coworkers [40] have also discussed the ^{19}F spectrum of per-fluorocyclooctane (VII), a compound which was first investigated by Thomas.[54] The ^{19}F spectrum of VII changes from a single line at room temperature to an approximate AB system and a single line at about −90 °C. No further change takes place down to −170 °C. At intermediate temperatures (about −65 °C), the central components of the AB system merge into one broad line, but the single line is still separate, although somewhat broadened also.

A boat-boat (or a twist-boat pseudorotating rapidly through the boat-boat) has been suggested as the conformation of VII.[40] The spectrum at low temperatures expected from the boat-boat is the sum of an AB and A_2 system of equal intensity, in agreement with experiment. However, attempts to calculate the line shape at intermediate temperatures, with the assump-

F. A. L. Anet

tion of a boat transition state were not successful.[40] It was suggested that the interconversion mechanism is interconversion to a boat-chair, which pseudorotates rapidly. Line shape calculations [55], however, do not fit this model either, nor a model where the boat-chair both pseudorotates and inverts rapidly.

The spectrum actually requires that exchange within the AB quartet be much faster than exchange with the A_2 system and this can only be done by having two separate processes. In one of these processes, the only exchange is within the AB quartet as in the following pseudorotation sequence: BB → BC → TC → BC → BB. In this model the boat-chair must not pseudorotate rapidly, nor must the twist-chair interconvert rapidly with the chair. The process, which is a pure ring inversion of the boat-boat, does not appear very probable because of the restrictions stated, but the presence of many CF_2 groups could conceivably cause unusual effects. To complete the explanation of the spectral features at intermediate temperatures requires a second process, *e.g.* pseudorotation through the boat, or some leakage of the twist-chair to the chair. The system now has enough freedom with two arbitrarily different rate constants that *any* exchange scheme of an AB plus A_2 system can be simulated.

Since the ^{19}F spectra of 1,1,2,2-tetrafluorocyclooctane can be explained satisfactorily on the basis of two different conformations, one in the crown, and the other in the boat-chair family, the question arises as to whether the spectra of perfluorocyclooctane can be explained on a similar model. The answer is that such an explanation is possible, but only if some unusual effects are accepted. The AB quartet at −65 °C can be assigned to a crown family conformation which is inverting slowly but pseudorotating rapidly. The single line would then have to be a rapidly inverting and pseudorotating boat-chair. The problem with this picture is that the single line remains unchanged to −170 °C, and this requires the boat-chair to invert rapidly at this temperature. A very low barrier to ring inversion in the boat-chair does not occur in the previously discussed compounds. Because perfluorocyclooctane must be a very strained molecule, however, analogies with simple cyclooctane derivatives may not be valid. It seems clear that more evidence is needed in order to resolve the conformational picture in perfluorocyclooctane.

In the discussion on 1,1,4,4-tetrafluorocyclooctane the argument was made that substitution of a methylene group in cyclooctane by a difluoromethylene group increases the barrier to pseudorotation by about 1.2 kcal/mole, owing to the presence of additional non-bonded repulsions in the transition state for the boat-chair to twist-boat-chair interconversion. Supporting evidence for this view comes from proton nmr studies on 1,1-dimethylcyclooctane (VIII) and on the ethylene ketal (IX) and the ethylene dithioketal (X) of specifically deuterated cyclooctanone.

202

IX   X

Table 8. Barriers in cyclooctane and some derivatives

Substituents on cyclooctane	$\Delta G^{\ddagger}$ (kcal/mole)		Ref.
	Ring inversion	Pseudo-rotation	
None	8.1	<5	47)
Methyl	ca. 8	<5	46)
1,1-Difluoro	7.5	4.9	40)
1,1-Ethylenedioxy	7.6	5.3	56,57,58)
1,1,4,4-Tetrafluoro	1)	6.1	40)
1,1-Ethylenedithioxy	8.5	6.6	38,57)
1,1-Dimethyl	8.0	8.0	38,59)
1,2,5,6-Tetrabromo	11.1	—	28)

1) The ring inversion barrier is difficult to obtain because of a nearly degenerate spectrum.

The dithioketal (X) gives particularly clean-cut results and will be discussed first. The γ-proton label in X gives rise to a single line at room temperature, and to two equal-intensity lines with a chemical shift difference of 0.42 ppm below −110 °C. Below about −140 °C the high-field component of the doublet splits into two lines separated by 0.22 ppm, while the low field component also gives two lines but separated by only 0.06 ppm. Thus, there are two processes which affect the spectra of X, and free energies of activation are given in Table 8.

Compound X is actually a racemic mixture because of the γ-proton label, but in the following discussion we will refer only to one of the two enantiomers. The nmr results can be interpreted in terms of the following four conformations.

The interconversion of the BC-2,5e with the BC-8,3e (or the BC-8,5a and BC-2,7a) can be accomplished by two different pseudorotation itineraries, namely, by passage through the BC-1 or BC-5 conformations. Irrespective of the path, this interconversion has the interesting effect of shifting the γ-proton from one axial site to another axial site (or one equatorial site to another equatorial site). This is consistent with the large chemical shift

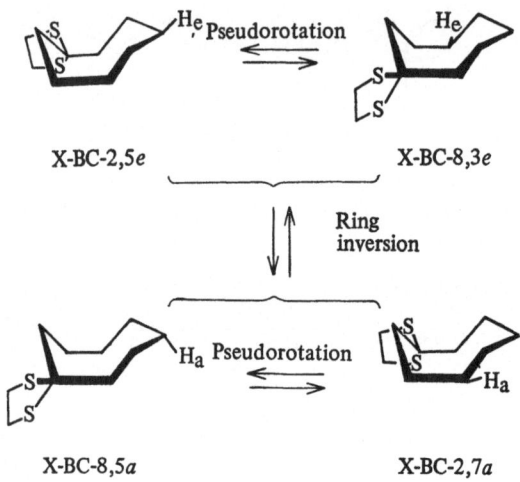

splitting (0.42 ppm) observed at intermediate temperatures, and the much smaller additional splitting found at very low temperatures. In contrast to 1,1-difluorocyclooctane, there is no evidence for a BC-5 conformation being appreciably populated in the dithioketal X.

The ethylene ketal, IX, also shows two distinct nmr processes, but the $\Delta G^{\neq}$ for pseudorotation is distinctly lower than in X (see Table 8). Rather broad spectra were obtained at the temperatures (ca $-173\,°C$) required to slow down pseudorotation in IX, and nothing can be said about the possible presence of the BC-5 form. Apart from this, the spectra of IX are quite analogous to those of X.

1,1-Dimethylcyclooctane (VIII) shows only a single observable change at 100 MHz in its proton spectrum. The methyl band, which is a single line at room temperature becomes a doublet below $-120\,°C$. There are also changes in the ring proton bands at about the same temperature. The spectrum does not show any further changes below $-130\,°C$. Because of the complexity of the ring proton bands, it is not possible to determine whether pseudorotation is slow or rapid at $-130\,°C$. In contrast, ^{13}C Fourier transform spectra, obtained with protons noise decoupled, (Fig. 5), are strikingly simple and informative. The methyl ^{13}C band splits in two below $-120\,°C$, and so do all the ring carbons, with the exception of C-1 and C-5. The spectrum at $-130\,°C$ corresponds exactly to a conformation which lacks any element of symmetry, $e.g.$ BC-2 and its mirror-image, BC-8. Therefore, pseudorotation and ring inversion must both be slow and from the line shapes at intermediate temperatures, it is apparent that $\Delta G^{\neq}$ for both processes is about 8.0 kcal/mole at $-120\,°C$.

The pseudorotation barriers given in Table 8 are strongly dependent on the size of the CX_2 group, whereas the barrier to ring inversion remains

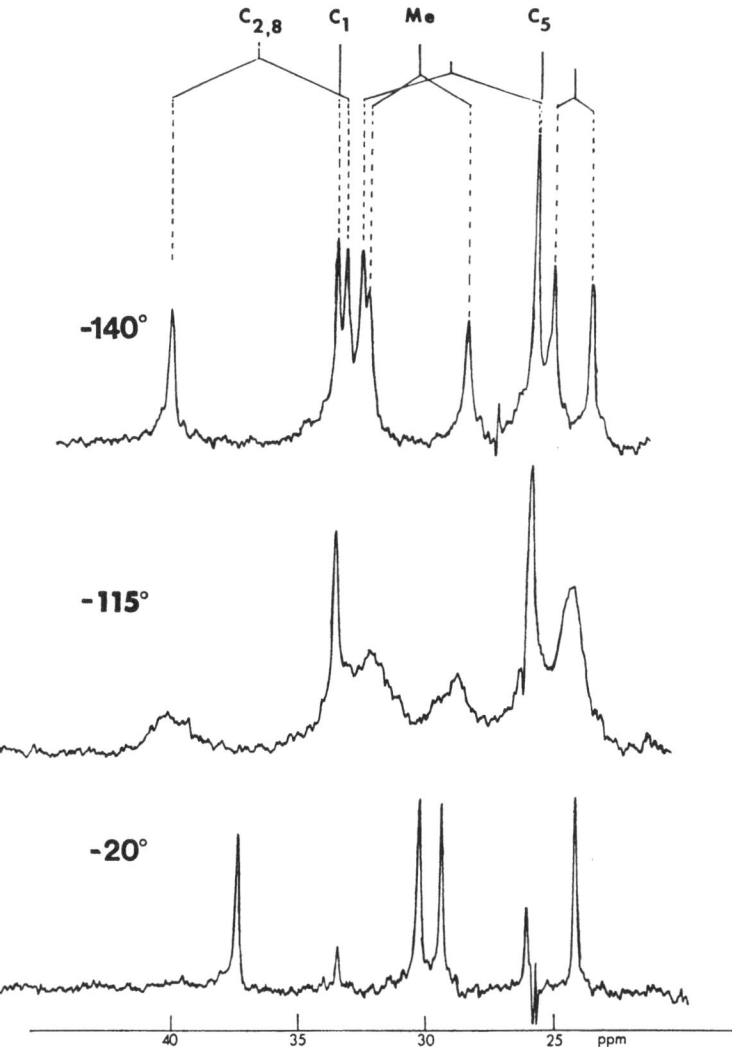

Fig. 5. Fourier transform ^{13}C spectra of 1,1-dimethylcyclooctane at various temperatures.[59] Protons are noise decoupled. Only a partial assignment of carbon resonances has been made

more or less constant at 8.0 ± 0.5 kcal/mole. For the 1,1-dimethyl compound (VIII), the BC-1 or the BC-3 and BC-7 conformations must be intermediates in the pseudorotation itinerary. Hendrickson's calculations (Table 3) for the excess strain energy of a methyl group show that these intermediates must be about 7 to 8 kcal/mole higher in energy than the BC-2 or BC-8 conformation, and this is consistent with a barrier to pseudorotation of 8 kcal/mole in VIII.

Ring inversion through the chair form can take place with the methyl groups in VIII remaining in relatively unhindered positions during the entire process, and thus the barrier to ring inversion is not expected to be greatly affected by the presence of two geminal substituents, in agreement with the data in Table 8.

The final compound to be considered in this section is *trans-syn-trans*-1,2,5,6-tetrabromocyclooctane [28] (XI), which exists in the crystalline state as the twist-chair-chair (TCC-1a,2a,5e,6e). Nmr data are in agreement with this conformation in solution, and the barrier to ring inversion is 11.1 kcal/mole at $-66\,°C$. This is a much higher barrier than is found in the cyclooctane boat-chairs, and agrees with similar barriers found in crown family conformations of heterocyclic eight-membered rings. The nmr spectra apparently indicate the presence of 10% of another conformation, which is probably the boat-chair, BC-1e,2e,5e,6e. The boat-chair has all the bromines equatorial, whereas the twist-chair-chair has one pair of bromines di-axial and the other pair di-equatorial. Di-equatorial bromines, especially if the external torsion angle (Br-C-C-Br) is smaller than 60°, are unfavorable because of repulsions between the bromine atoms. This could be the reason for the tetrabromo compound existing predominantly in the twist-chair-chair. The results clearly show that the twist-chair-chair cannot be very much higher in energy than the boat-chair for cyclooctane itself, otherwise the substituent effects would not be able to make the twist-chair-chair the major conformation in XI.

XI-TCC-1a,2a,5e,6e XI-BC-1e,2e',5e,6e

Conformational energy barriers have been studied in various cyclooctane derivatives by mechanical relaxation methods.[60,61] The frequency and temperature range of such measurements are very large, but the identities of the processes observed are not as clear as in nmr measurements. With poly(cyclooctyl methacrylate) a process with an activation energy of 10.6 kcal/mole is found, and has been interpreted in terms of a boat-chair ring inversion.[60] However, nmr results on various cyclooctane derivatives, including cyclooctyl formate and cyclooctyl acetate [56], show that 10.6 kcal/mole is too high a value for ring inversion. A more likely process, in our view, is a boat-chair to twist-chair-chair interconversion, which is expected to have a slightly higher barrier than the boat-chair ring inversion.

In summary, the experimental nmr data presented in this Section stongly support the boat-chair as the lowest energy conformation for simple cyclooctane derivatives. The twist-chair-chair is of next lowest energy and the presence of certain substituents can make this conformation be the dominant one. Boat-boat family conformations are only (if ever) found in very special compounds.

B. Cyclooctanone and Related Compounds

The 60 MHz spectra of the cyclooctanone-d_{13} isomers, XII and XIII are extremely informative.[38,56,57,58] The isomer XII which has a single proton in the γ position, gives a single line at room temperature, two lines below $-122\,°C$, and four lines of equal intensity below $-147\,°C$. Isomer XIII, which has the proton label in the δ position gives a single line at room temperature, and two lines below $-112\,°C$, with no further change at lower

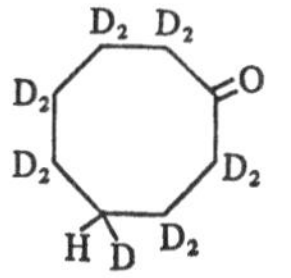

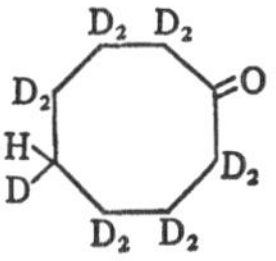

temperature. The high-field line in the doublet of XIII is only 0.63 ppm downfield from internal tetramethylsilane, some 0.5 ppm more shielded than expected for a methylene proton.

The γ labelled isomer XII is actually a racemate, but it is sufficient to consider one of the enantiomers as was done with the ethylene dithioketal of XII, which was discussed in Section VII. A. There are then four conformations: BC-3,6a, BC-7,2e', BC-3,8a', and BC-7,4e, as shown below.

The interconversion of BC-3,6a with BC-7,2e' on the one hand and BC-3,8a' with BC-7,4e on the other hand can occur by two different pseudorotation itineraries, and evidence will be presented later that the itinerary in which the BC-1 form is an intermediate is of lower energy than the alternative itinerary, in which the BC-5 form is an intermediate. The pseudorotation just described accounts for the two lines at intermediate temperatures

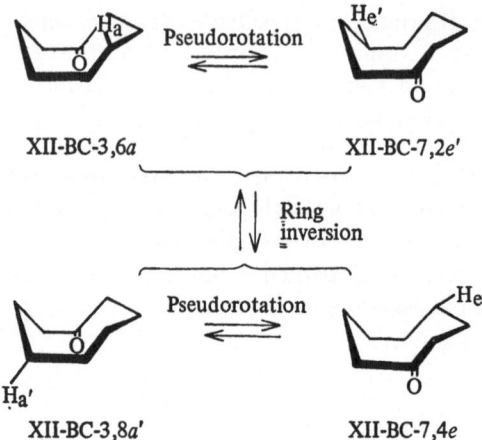

in the γ isomer. In the δ isomer this process has no effect on the δ proton chemical shifts as it results in an interconversion of the type BC-3,7$a \rightleftarrows$ BC-7,3a. These results show that the boat to twist-boat-chair interconversion is indeed the process with the lower barrier in XII. The higher temperature process in XII and the only process observed in XIII are then ring inversion.

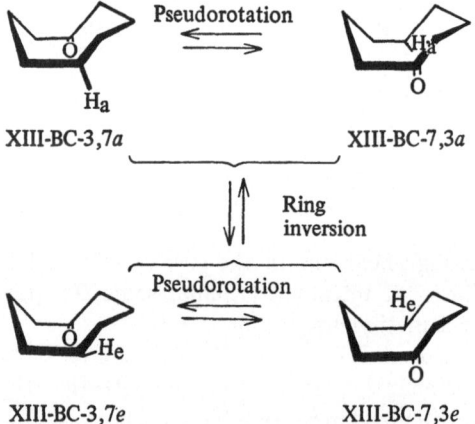

The high-field chemical shift observed in XIII at low temperatures is well explained by the boat-chair conformation, since the δ proton in the axial position is directly above the π bond of the carbonyl group and should thus be strongly shielded.[5]

Table 9 gives the free energy barriers for conformational interconversions in cyclooctanone. This table also gives the barrier to pseudorotation in

Table 9. Barriers in cyclooctanone and related compounds

Compound	$\Delta G^{\neq}$ (kcal/mole)		Ref.
	Ring inversion	Pseudo-rotation	
Cyclooctanone	7.5	6.3	[38,56,57,58]
5-t-Butylcyclooctanone	—	8.0	[56,58]
Methylenecyclooctane	8.1	—	[38,57]

5-*t*-butylcyclooctanone (XIV). No ring inversion is to be expected in XIV since the large *t*-butyl group must take up the equatorial position exclusively, and thus the one nmr process observed in the ring proton bands must be pseudorotation. The abnormally large barrier to pseudorotation in XIV finds a ready rationalization in the fact that the pseudorotation itinerary in XIV, unlike that of cyclooctanone, must proceed through the BC-5 conformation (where 5 indicates the carbonyl position) in order for the *t*-butyl group to remain equatorial during the entire itinerary. This also indicates that cyclooctanone itself does not pseudorotate most easily *via* the BC-5 conformation, as otherwise there should be little difference in the pseudorotation barriers in the two compounds.

The BC-3 conformation for cyclooctanone is supported by recent strain energy calculations, which have already been mentioned (Section V). Qualitatively, the BC-3 conformation is also very reasonable, since the non-bonded repulsions between the 3 and 7 methylene groups in the cyclooctane boat-chair conformation are largely removed in the BC-3 form. The 1 position in the boat-chair also has the same kind of advantage that the 3 position has. However, the 3 position is also favored because of the relief of eclipsing strain which occurs in that position, but not in the 1 position (see Table 2 for dihedral angles in the boat-chair). This point will be amplified in the following discussion on methylenecyclooctane.

The nmr spectra of the methylenecyclooctanes XV and XVI corresponding to ketones XII and XIII show the following features: [38,56,57] only a single process is observed, and the δ proton in XVI does not have an unusually shielded chemical shift. The barrier for this process (Table 9) corresponds to a ring inversion. The evidence strongly suggests the symmetrical BC-1 conformation for methylenecyclooctane. The BC-1 form immediately explains the absence of a pseudorotation process, since such a process is invisible (although by no means non-existent) in that conformation. The normal chemical shifts of the axial δ proton is also consistent with the BC-1 conformation, since that proton is not near the face of the vinylidene group, as it is in the BC-3 form.

XV-BC-1,4a   XV-BC-1,6e

XVI-BC-1,5a   XVI-BC-1,5e

It is perhaps surprising that methylenecyclooctane does not have an analogous conformation to that of cyclooctanone. A possible explanation is based on the fact that the carbonyl and vinylidene groups are quite different from one another where eclipsing effects are concerned. For example, the barriers to methyl rotation in acetone and isobutylene are 0.8 and 2.2 kcal/mole, respectively.[62,63] This means that, other things being equal (e.g. non-bonded repulsions), the carbonyl group should prefer a position which, in the corresponding hydrocarbon, is as much eclipsed as possible. The vinylidine group, on the other hand should not show a strong preference for eclipsed positions. The energy difference from this effect is of the order of 1.4 kcal/mole, which is quite sufficient to explain the nmr results, especially since the significant data refers to temperatures below −100 °C. At room temperature, the BC-3 conformation should still be dominant for cyclooctanone, but a small percentage of the BC-1 conformation may well occur. Conversely, methylenecyclooctane should be mainly in the BC-1 conformation at room temperature, with a minor amount of the BC-3 form also being present. Unfortunately, nmr becomes a very uncertain tool when applied to systems which are undergoing rapid averaging, even though, in principle, information could be obtained from the temperature dependence of chemical shifts. A safer approach is to look for minor conformations at low temperatures, and the best method is probably Fourier transform ^{13}C nmr.[50,64]

C. Heterocyclic Eight-Membered Rings

The conformational picture in heterocyclic eight-membered rings is quite varied, even though only a few classes of compounds have been investigated. We will begin with heterocycles containing up to four oxygen atoms in the ring.

Oxocyclooctane, or oxocane (XVII), has been studied by Anet and De-gen.[65] Although the 60 MHz proton spectrum of XVII shows little change at low temperatures, a single process is easily observed at 251 MHz. The α protons, for example, give one broad line at room temperature and two broad lines below − 122 °C, with the rather small chemical shift difference of 0.18 ppm. No further change takes place at lower temperatures, and the ^{13}C nmr spectrum of XVII is temperature independent to − 170 °C.

Since oxygen is much smaller than a methylene group, the same kind of situation occurs in XVII as was discussed in the previous section. The barrier to methyl rotation in dimethyl ether is 2.7 kcal/mole [66], only slightly lower than in propane, where the barrier is 3.4 kcal/mole. Oxocane should there-fore have the BC-1 conformation, as in methylenecyclooctane rather than the BC-3 and BC-7 conformations. The presence of only a single process in the proton spectrum of XVII is immediately consistent with the BC-1 conformation, but requires rapid pseudorotation between the BC-3 and BC-7 forms at − 170 °C if the latter two forms are the correct conformations. The pseudorotation barrier in XVII should be higher than in cyclooctane, and probably comparable to that in cyclooctanone (6.3 kcal/mole). Thus, pseudorotation of the BC-3 form should not be rapid at − 170 °C, and further support for this hypothesis is provided by 1,3-dioxocane (see below). It is therefore probable that oxocane has the BC-1 conformation.

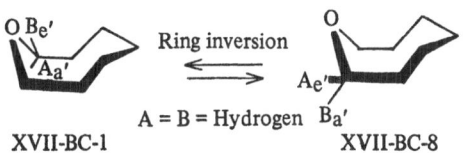

A = B = Hydrogen

XVII-BC-1   XVII-BC-8

1,3-Oxocane (XVIII) and several of its *gem*-dimethyl derivatives have been studied by nmr.[65,67]

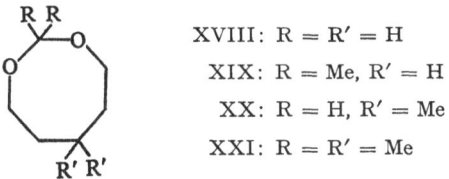

XVIII: R = R′ = H
XIX: R = Me, R′ = H
XX: R = H, R′ = Me
XXI: R = R′ = Me

The methylenedioxy proton band in XVIII changes from a single line at room temperature to an AB quartet ($\delta_{AB} = 0.13$ ppm) below − 125 °C. No further change takes place in these bands down to − 170 °C, but the other methylene protons give complex bands which show two clear nmr processes, one at about − 125 °C, and the other at − 155 °C.

The nmr data can be simply explained on the basis of the BC-1,3 and its mirror image BC-1,7. Below − 155 °C both pseudorotation and ring inversion are slow; above − 155 °C, pseudorotation becomes fast and above − 125 °C ring inversion is also fast. The BC-1,3 and BC-1,7 conformations have oxygens at the favorable 1,3 and 7 positions, and furthermore these forms have the ether dipoles in a low-energy arrangement, exactly as found in dimethoxymethane.[68] The measured dipole moment of 1,3-dioxocane is 0.7 D, in agreement with the BC-1,3 form.[67] Free energy barriers for 1,3-dioxocane are given in Table 10.[65]

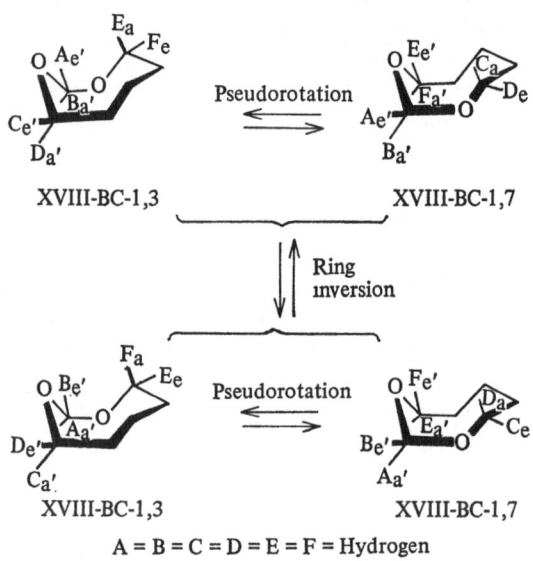

A = B = C = D = E = F = Hydrogen

The isomeric dimethyl-1,3-dioxocanes, XIX and XX, have coalescence temperatures of − 80 °C and − 120 °C respectively, while the 2,2,6,6-tetramethyl compound XXI has a coalescence temperature of − 70 °C.[67] Boat-chair conformations (BC-1,3 and BC-1,7) have been assigned to the 2,2-dimethyl derivative XIX. For the 6,6-dimethyl and the 2,2,6,6-tetramethyl derivatives, boat-boat conformations were suggested, since the boat-chair with an axial methyl group in the 4 or 6 position was considered to be impossibly strained on the basis of Hendrickson's calculations (Table 3). However, the situation here is not nearly as bad as in the corresponding cyclooctane system. In the BC-1,3 conformation the axial methyl group at the 6 position should have a much smaller repulsive interaction with the 1-oxygen than with a methylene group in the corresponding dimethyl-cyclooctane. Further nmr work on these compounds, carried out at 251 MHz,

Table 10. Barriers in oxocanes

Compound	$\Delta G^{\ddagger}$ (kcal/mole)[1]	
	Ring inversion	Pseudo-rotation
Oxocane	7.4	
1,3-Dioxocane	7.3	5.7
1,3,6-Trioxocane	8.7	
	6.8	
1,3,5,7-Tetroxocane	12.9	—

[1] Data from Ref.[65].

has revealed a second nmr process in each case.[69] The conformations of XX and XXI therefore cannot have any symmetry and the boat-boats, which have C_2 symmetry in the above compounds, can be excluded. Twist-boats do not appear likely because the barrier to pseudorotation through the boat-boat should be even less than in cyclooctane, where calculations give a barrier of only 0.5 kcal/mole. Thus the twist-boats and boat-boats should give the same kind of spectra, and both can be excluded. It appears that boat-chairs offer satisfactory explanations of the nmr spectra of XX and XXI [69], although some unusual features are found. In XIX and XXI, the time-average symmetry at intermediate temperatures is C_2, rather than the normal C_s symmetry, as is found in XX or 1,3-dioxocane itself. This must mean that the boat-chair to twist-boat-chair interconversion is the process with the higher energy barrier, just the reverse of the usual situation. To be consistent, we have to define the high-energy process as ring inversion. The low-energy process, which is responsible for the C_2 time-average symmetry at intermediate temperatures, is probably a pseudo-rotation *via* the twist-boat-chair, the twist-boat and the boat-boat forms. In Section VI this interconversion was suggested as a mechanism for ring inversion but if the boat-chair to twist-boat-chair interconversion is slow, this interconversion is no longer a ring inversion, but is then simply a pseudo-rotation. Further experiments are needed to clarify the conformational picture in these methylated dioxocanes. It is clear, however, that *gem*-dimethyl groups drastically increase the barrier to the boat-chair to twist-boat-chair interconversion, just as in 1,1-dimethylcyclooctane (Section VII.A).

The cyclic formal of diethylene glycol, *i.e.* 1,3,6-trioxocane, shows a very interesting temperature-dependent nmr spectrum, especially at 251 MHz.[65,67] The C-2 proton band changes from a single line at room temperature to a widely-spaced AB quartet ($\delta_{AB} = 0.8$ ppm) and a single line below $-80\,°C$. The intensities of the AB quartet and the single line are approximately the same. At still lower temperatures, only the single line changes, giving rise to a narrowly-spaced AB quartet ($\delta_{AB} = 0.07$ ppm) below $-135\,°C.$[65] The widely-spaced AB quartet has been assigned to a crown family conformation and the narrowly-spaced quartet to a boat-chair. The chemical shift difference between the protons on C-2 in 1,3-dioxocane, which exists as a boat-chair, is also quite small, namely 0.13 ppm, and this should be a general feature of methylene protons at the 2 and 8 positions in the boat-chair, because these carbon atoms lie on local and approximately C_2 axes. The most likely boat-chair for trioxocane is the BC-1,3,6 conformation.

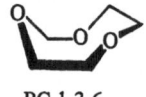

BC-1,3,6

The final compound in the oxygen series of heterocycles which has been studied by nmr is the cyclic tetramer of formaldehyde or 1,3,5,7-tetroxocane.[65,67] The proton spectrum is a single line at room temperature and changes to a widely-spaced AB quartet ($\delta_{AB} = 0.64$ ppm) and a single line below $-15\,°C$. No further splittings take place down to $-170\,°C$, although the single line is quite broad at $-160\,°C$, and might actually be an unresolved AB quartet with a very small chemical shift.[65]

The nmr spectra of tetroxocane bear a striking resemblance to those of trioxocane, and similar conclusions can be drawn. Thus, the widely-spaced AB quartet can be assigned to a crown family conformation. Because the cyclic tetramer of acetaldehyde has been found to have a crown conformation by X-ray diffraction [29], it is quite likely that tetroxocane exists in solution at least in part as a crown. The single line can then be assigned to a boat-chair family conformation, which is pseudorotating rapidly, and may or may not also be inverting rapidly depending on whether the "single line" really represents a single chemical shift or an unresolved AB quartet. An interesting feature of the nmr spectra of tetroxocane is the dramatic temperature-dependence in the relative population of the AB quartet and the single line.

Thermodynamic calculations, based on integrated intensity measurements, show that the crown form has the lower enthalpy, and also the lower entropy ($\Delta S = 8.5$ [67] or 6 ± 2 eu [65]). The low entropy of the crown finds an explanation in the high symmetry of this conformation, compared to the low symmetry of the boat-chair. The best boat-chair for tetroxocane appears to be the BC-1,3,5,7 conformation. However, the ether functions at positions 3 and 7 have their dipoles close together in an unfavorable orientation but a slight distortion of the boat-chair in the direction of the twist-boat-chair should relieve this repulsion.

The appearance of crown forms in tetroxocane and other compounds with heteroatoms in the 1,3,5 and 7 positions probably results from two stabilizing effects: (a) a slightly lower eclipsing barrier for the —O—CH$_2$— versus the —CH$_2$—CH$_2$ fragment; in the crown there are eight partially eclipsed bonds which would benefit from relief of torsional strain; (b) distorted crowns, *i.e.* the chair-chair and twist-chair-chair may suffer from higher dipolar repulsions than does the crown. Unfortunately, there are no conformational energy calculations available on any of the heterocyclic systems.

The sulfur analogue of tetroxocane, namely 1,3,5,7-tetrathiocane, gives rise to a single line in the proton spectrum from room temperature down to —170 °C.[27] Since crown family conformations have invariably been found to have relatively high barriers to inversion and large chemical shift differences between geminal protons, a crown conformation seems to be excluded for tetrathiocane. The structure of tetrathiocane in the crystalline state is in fact the boat-chair, BC-2,4,6,8.[27] This conformation has somewhat different dihedral angles from the regular cyclooctane boat-chair, and the relatively long carbon-sulfur bonds allow the methylene groups to take up positions which would normally have larger non-bonded repulsions. In contrast, the BC-2,4,6,8 conformation of tetroxocane appears from molecular models to have a large amount of non-bonded repulsions as a result of the comparatively short carbon-oxygen bonds.

The data on barriers to ring inversion in heterocyclic boat-chairs (Table 10) show a trend to lower barriers when the number of hetero-atoms is increased, and in the absence of *gem*-dimethyl substitution, pseudorotation barriers are also quite low. It is therefore reasonable to expect that the barriers to pseudorotation and inversion in the tetrathiocane boat-chair might be too low for nmr detection, and this would, of course, explain the temperature-independent spectrum.

A series of eight-membered rings with various heteroatoms at the 1,3,5 and 7 positions has been studied by Lehn and Riddell.[70] Compound XXII (R = Me; X = S) (Table 1) is known to be a crown in the crystalline state [30], and the dynamic nmr spectra are also consistent with this conformation. The free energy barrier to ring inversion (14.8 kcal/mole) and the chemical shift difference between geminal protons (0.32 ppm) are both much larger than would be expected for a boat-chair, but are of the order of magnitude found in other crowns. Barriers to ring inversion in related compounds, where the nitrogen substituent is different from methyl, or where selenium replaces sulfur, are given in Table 11. Since all the barriers are relatively high, it appears very likely that these compounds also exist in crown conformations.

XXII

Table 11. Barriers in diazadithio and
diazadiselenia cyclooctanes

| Groups in XXII | | $\Delta G^{\pm}$ (kcal/mole) |
R	X	
Me	S	14.8
Et	S	14.6
iPr	S	14.6
Ph	S	13.4
Me	Se	14.4

The final compound to be discussed in this section is 5-oxocanone, (XXIII), a molecule which has a potential transannular interaction of the ether oxygen with the carbonyl group.[71] The proton nmr spectrum shows two processes with free energies of activation of 9.0 and 7.8 kcal/mole. The ^{13}C nmr spectrum shows only a single process, which corresponds to the lower-energy process observed in the proton spectrum. The spectra are quite analogous to those of cyclooctanone (Section VII.B), and strongly support a conformation without any symmetry. The BC-3,7 conformation shown below represents one of two possible mirror-image conformations. This conformation fits the nmr data and is also expected to be the most stable boat-chair form of XXIII. As with cyclooctanone, the lower energy process has been ascribed to pseudorotation of the boat-chair *via* the twist-boat-chair.[72]

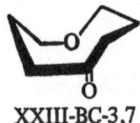

XXIII-BC-3,7

Since the interconversions of the conformations of XXIII are very similar to those described in detail for cyclooctanone, no further discussion will be given here.

VIII. Conclusions

Most "simple" cyclooctane compounds, including ketones and heterocyclic analogues, exist predominantly in boat-chair conformations. Compounds with special substitution patterns, or with hetero-atoms, often exist partially or mainly in crown family conformations (crowns or twist-chair-chairs). Boat-chairs generally have medium barriers (*ca.* 8 kcal/mole) to ring inversion, and, except in monosubstituted cyclooctanes and cyclooctane itself, have barriers to pseudorotation (*via* the twist-boat-chair) which depend a great deal on substituents, but which are large enough (>4.5 kcal/mole) for measurements by dynamic nmr methods. Cyclooctane itself has an unobservably low barrier to pseudorotation.

Ring inversion of crown family conformations generally proceeds by interconversion, over a relatively high barrier (10 to 12 kcal/mole), to boat-chairs, which invert relatively easily. Pseudorotation barriers in crown family conformations such as the twist-chair-chair are extremely low and not detectable by direct line-shape measurements.

The least known conformational family is the boat-boat. It seems that a very special substitution pattern is required to make conformations in this family more stable than those of the crown and boat-chair. Highly fluorinated cyclooctanes possibly exist in the twist-boat or boat-boat conformations; however, additional evidence is badly needed to support the conformational assignments which have been made. A boat-boat conformation (in the cyclooctane nomenclature) does occur in [3.3.1]bicyclononane, which contains a 1,5-bridged cyclooctane ring.[73]

While only further experiments will show whether the conformational picture outlined in this chapter has real validity, the author believes that the main conclusions drawn will stand the test of time.

Added in proof. The conclusions reached in Section VII. A concerning the probable presence of a small amount of a crown-family conformation in cyclooctane has now been directly verified by both ^{13}C and ^{1}H nmr spectroscopy.[74] The crown-family conformation occurs at concentrations of 6, 2 and 0.3% at room temperature, −45° and −125°C 125°C respectively. The $\Delta H°$ and $\Delta S°$ for the boat-chair to crown conformation are 1.9 ± 0.2 kcal/mole and 1 ± 1 eu respectively and the $\Delta G^{\pm}$ for this process is 11.2 ± 0.4 kcal/mole at −45°.

The presence of about 10% boat-chair conformation in *trans-syn-trans* 1,2,5,6-tetrabromocyclooctane (XI) (see Section VII. C) has been confirmed by 251 MHz ^{1}H nmr.[75] The *trans-anti-trans* isomer of this compound has been found by 251 MHz ^{1}H nmr to be an approximately 1:1 mixture of a crown-family and a boat-chair conformation.[75]

Boat-chairs have been assigned to *cis-* and *trans-*1,5-diacetoxycylooctanes and the former shown to undergo pseudorotation with a barrier ($\Delta G^{\pm}$) of 6.5 kcal/mole (the ring inverted form is present in too small a concentration to be detected).[76] The *trans* isomer has a barrier to pseudorotation which is too low for nmr measurement, but the barrier to ring inversion (8.0 kcal/mole) can now be obtained because the conformations which are related by inversion are equally populated. The barriers are similar to those in compound IX, and the interconversion scheme given [76] is in complete agreement with the principles given in this chapter.

F. A. L. Anet

1,5-Cyclooctadione and 1,5-dimethylenecyclooctane both appear to exist in BC-3,7 conformations.[77] 1,2-Dithiacyclooctane exists in the BC-3,4 form with a barrier to ring inversion of 9.1 kcal/mole, and an unobservably low barrier to pseudorotation.[78] This latter process (BC-3,4 $\rightleftharpoons$ BC-6,7) does *not* require passage through a 0° torsional angle for the disulfide group. The barrier for the boat-chair pseudorotation in cyclooctane has been calculated to bee 3.3 kcal/mole [79].

Acknowledgements. I wish to thank the National Science Foundation for financial support of our work on the conformations of eight-membered rings, and I am most grateful to my collaborators, Drs. S. Hartman, M. St. Jacques, G. N. Chmurny, P. M. Henrichs, W. D. Larson, J. J. Wagner, P. J. Degen and J. E. Krane.

References

1) Pople, J. A., Schneider, W. G., Bernstein, H. J.: High-resolution nuclear magnetic resonance. New York: McGraw-Hill 1959. — Bovey, F. A.: Nuclear magnetic resonance spectroscopy. New York: Academic Press 1969. — Becker, E. D.: High resolution NMR. New York: Academic Press 1969.
2) Johnson, C. S., Jr.: Advan. Magn. Resonance *1*, 33 (1965).
3) Reeves, L. W.: Advan. Phys. Org. Chem. *3*, 187 (1965).
4) Binsch, G.: Top. Stereochem. *3*, 97 (1968).
5) Anet, F. A. L., Anet, R.: Det. Org. Struct. Phys. Methods *3*, 344 (1971).
6) Eliel, E. L., Allinger, N. L., Angyal, S. J., Morrison, G. A.: Conformational Analysis. New York: Wiley 1965.
7) Lambert, J. B., Oliver, W. L., Jr.: Tetrahedron Letters *1968*, 6187.
8) Kilpatrick, J. E., Pitzer, K. S., Spitzer, R.: J. Am. Chem. Soc. *69*, 2483 (1947). — Pitzer, K. S., Donath, W. E.: J. Am. Chem. Soc. *81*, 3213 (1959).
9) Berry, R. S.: J. Chem. Phys. *32*, 933 (1960).
10) Hendrickson, J. B.: J. Am. Chem. Soc. *86*, 4854 (1964).
11) Hendrickson, J. B.: J. Am. Chem. Soc. *89*, 7036 (1967).
12) Hendrickson, J. B.: J. Am. Chem. Soc. *89*, 7043 (1967).
13) Hendrickson, J. B.: J. Am. Chem. Soc. *89*, 7047 (1967).
14) Bellis, H. E., Slowinski, E. J., Jr.: Spectrochim. Acta *15*, 1103 (1959).
15) Sands, D. E., Day, V. W.: Acta. Cryst. *19*, 278 (1965).
16) Almenningen, A., Bastiansen, O., Jensen, H.: Acta Chem. Scand. *20*, 2689 (1966).
17) Anet, F. A. L., St. Jacques, M.: J. Am. Chem. Soc. *88*, 2585 (1966).
18) Dowd, P., Dyke, T., Neumann, R. M., Klemperer, W.: J. Am. Chem. Soc. *92*, 6325 (1970).
19) Dobler, M., Dunitz, J. D., Mugnoli, A.: Helv. Chim. Acta *49*, 2492 (1966).
20) Bürgi, H. B., Dunitz, J. D.: Helv. Chim. Acta *51*, 1514 (1968).
21) v. Egmond, J., Romers, C.: Tetrahedron *25*, 2693 (1969).
22) Groth, P.: Acta Chem. Scand.: *19*, 1497 (1965); Acta Chem. Scand. *21*, 2695 (1967).
23) Srikrishnan, T., Srinivasan, R., Zand, R.: J. Cryst. Mol. Struct. *1*, 199 (1971); Srinivasan, R., Srikrishnan, T.: Tetrahedron *27*, 1009 (1971).
24) Johnson, S. M., Maier, C. A., Paul, I. C.: J. Chem. Soc. (B) *1970*, 1603.
25) Go, K. T., Paul, I. C.: Tetrahedron Letters, *1965*, 4265. — Paul, I. C., Go, K. T.: J. Chem. Soc. (B) *1969*, 33.
26) Dunitz, J. D.: Pure Appl. Chem. *25*, 495 (1971).
27) Frank, G. W., Degen, P. J., Anet, F. A. L.: J. Am. Chem. Soc. *94*, 4792 (1972).

28) Ferguson, M. N., MacNicol, D. D., Oberhausli, R., Raphael, R. A., Zabkiewiez, R.: Chem. Commun. *1968*, 103.
29) Pauling, L., Carpenter, D. L.: J. Am. Chem. Soc. *58*, 1274 (1937).
30) Grandjean, D., Leclaire, A.: Compt. Rend. *265*, 795 (1967).
31) Cady, H. H., Larson, A. C., Comer, D. T.: Acta Cryst. *16*, 617 (1963).
32) Schenck, H.: Acta Cryst. *B27*, 185 (1971).
33) Sass, R., Donohue, J.: Acta Cryst. *11*, 497 (1958). — Lund, E. W., Svendsen, S. R.: Acta Chem. Scand. *11*, 940 (1957).
34) Abrahams, S. C.: Acta Cryst. *8*, 661 (1955).
35) Wiberg, K. B.: J. Am. Chem. Soc. *87*, 1070 (1965).
36) Bixon, M., Lifson, S.: Tetrahedron *23*, 769 (1967).
37) Allinger, N. L., Hirsch, J. A., Miller, M., Tyminski, J. J., Van Catledge, F. A.: J. Am. Chem. Soc. *90*, 1199 (1968).
38) Henrichs, P. M.: Ph. D. Thesis, University of California, Los Angeles, 1969.
39) Program written by C. K. Johnson, Oak Ridge National Laboratory. Program and documentation available from Clearinghouse for Federal Scientific and Technical Information, National Bureau of Standards, U.S. Department of Commerce, Springfield, Virginia 22151.
40) Anderson, J. E., Glazer, E. D., Griffith, D. L., Knorr, R., Roberts, J. D.: J. Am. Chem. Soc. *91*, 1386 (1969).
41) Allinger, N. L., Greenberg, S.: J. Am. Chem. Soc. *81*, 5733 (1959).
42) Allinger, N. L., Jindal, S. P., Da Rooge, M. A.: J. Org. Chem. *27*, 4290 (1962).
43) Allinger, N. L., Tribble, M. T., Miller, M. A.: Tetrahedron *28*, 1173 (1972).
44) Lifson, S., Warshel, A.: J. Chem. Phys. *49*, 5116 (1968).
45) Boyd, R. H.: J. Chem. Phys. *49*, 2574 (1968).
46) Anet, F. A. L., St. Jacques, M.: J. Am. Chem. Soc. *88*, 2586 (1966).
47) Anet, F. A. L., Hartman, J. S.: J. Am. Chem. Soc. *85*, 1204 (1963).
48) Meiboom, S.: Paper presented at the Symposium on High Resolution Nuclear Magnetic Resonance at Boulder, Colorado, July 1962.
49) Anet, F. A. L.: Unpublished.
50) Anet, F. A. L., Bradley, C. H., Buchanan, G. W.: J. Am. Chem. Soc. *93*, 258 (1971).
51) Anet, F. A. L., St. Jacques, M., Chmurny, G. N.: J. Am. Chem. Soc. *90*, 5243 (1968).
52) Roberts, J. D.: Abstracts, 19th National Organic Symposium of the American Chemical Society, Tempe, Arizona, June 1965.
53) Gerig, J. T., Roberts, J. D.: J. Am. Chem. Soc. *88*, 2791 (1966).
54) Peake, A., Wyer, J. A., Thomas, L. F.: Chem. Commun. 95 (1966).
55) Anet, F. A. L., Larson, W. D.: Unpublished.
56) St. Jacques, M.: Ph. D. Thesis, University of California, Los Angeles, 1967.
57) Anet, F. A. L., St. Jacques, M., Henrichs, P. M.: Intrasc. Chem. Rep. *4*, 251 (1970).
58) Anet, F. A. L.: Paper presented at the Brussels International Symposium, September 1969. Published in: Conformational analysis, scope and present limitations (ed. G. Chiurdoglu), p. 15. New York: Academic Press 1971. Abstracts of the 20th National Organic Symposium of the American Chemical Society, Burlington, Vermont, 1967.
59) Anet, F. A. L., Wagner, J. J.: Unpublished.
60) Heijboer, J.: J. Polymer Sci. *C16*, 3413 (1965). Physics of non-crystalline solids. Proceedings of the Intern. Conf. Delft, July 1964 (ed. J. A. Prins), p. 231. Amsterdam: Thesis, Leiden 1972.
61) Frosini, V., Magagnini, P., Butta, E., Baccaredda, M.: Kolloid-Z. *213*, 115 (1966).
62) Laurie, V. W.: J. Chem. Phys. *34*, 1516 (1961).
63) Swalen, J. D., Costain, C. C.: J. Chem. Phys. *31*, 1562 (1959).

F. A. L. Anet

[64] Levy, G. C., Nelson, G. L.: Carbon-13 nuclear magnetic resonance for organic chemists. New York: Wiley 1972.
[65] Anet, F. A. L., Degen, P. J.: J. Am. Chem. Soc. *94*, 1390 (1972).
[66] Aoki, K.: J. Chem. Soc. Japan, Pure Chem. Sect. *74*, 110 (1953).
[67] Dale, J., Krane, J.: J. Am. Chem. Soc. *94*, 1389 (1972).
[68] Astrup, E. E.: Acta Chem. Scand. *25*, 1494 (1971).
[69] Anet, F. A. L., Krane, J.: Unpublished.
[70] Lehn, J. M., Riddell, F. G.: Chem. Commun. *1966*, 803.
[71] Leonard, N. J., Milligan, J. W., Brown, T. L.: J. Am. Chem. Soc. *82*, 4075 (1960).
[72] Anet, F. A. L., Degen, P. J.: Tetrahedron Letters 1972, 3613
[73] Dobler, M., Dunitz, J. D.: Helv. Chim. Acta *47*, 695 (1964). — Brown, W. A. C., Martin, J., Sim, G. A.: J. Chem. Soc. *1965*, 1844.
[74] Anet, F. A. L., Basus, V. J.: J. Am Chem. Soc. *95*, 4424 (1973).
[75] Anet, F. A. L., Kozerski, L.: Unpublished.
[76] St-Jacques, M., Prud'homme, R.: J. Am. Chem. Soc. *94*, 6479 (1972).
[77] Anet, F. A. L., Krane, J., Wong, L.: Unpublished.
[78] Anet, F. A. L., Degen, P. J.: Unpublished.
[79] Anet, F. A. L., Krane, J.: Tetrahedron Letters, in press.

Errata

Nakajima, T.: Quantum chemistry of nonbenzenoid cyclic conjugated hydrocarbons. Topics Current Chem. *32*, 1 (1972).

The contents from the 13th line from the bottom to the 4th line from the bottom on p. 20 should read:

...energy differences: the triplet state and $^1B_{2g}$, $^1B_{1g}$ and $^1A_{1g}$ singlet states in order of increasing energy [13,15]. The lowest singlet can, in principle, interact vibronically with $^1B_{1g}$ and $^1A_{1g}$ singlets through the a_{2g} and b_{2g} deformations of the nuclei, respectively. Of these two possible types of nuclear deformation, the a_{2g} distortion corresponds to a symmetric C—C stretching mode (a breathing mode), while the b_{2g} distortion corresponds to an antisymmetric C—C stretching mode (a bond alternation). The energy of the $^1B_{2g}$ state is thus lowered by the vibronic interaction with these states to such...

Professor Takeshi Nakajima, Sendai, Japan

Topics in Current Chemistry

Fortschritte der chemischen Forschung

Managing Editor:
F. Boschke

Volume 38

K. Dimroth: Phosphorus-Carbon Double Bonds

44 figures. III, 147 pages. 1973
DM 48,—; US $19.70 ISBN 3-540-06164-9

Prices are subject to change without notice

Contents: Delocalized Phosphorus — Carbon Double Bonds. — Phosphamethin-cyanines. λ^3-Phosphorins and λ^5-Phosphorins: Phosphamethin-cyanines. — λ^3-Phosphorins. — λ^5-Phosphorins. — Outlook.

The chemistry of phosphorus compounds with a delocalized P-C double bond proves to be very versatile.
λ^3-Phosphorins have physical properties which are rather similar to those of pyridines. But the chemistry of λ^3-phosphorins is very different, due mainly to the phosphorus atom which can easily lose one electron to produce a stable radical cation, or accept one or more electrons ro yield a radical anion, dianion or radical trianion.
Nucleophiles add to stable λ^4-phosphorin anions. No stable λ^4-phosphorinium compound could be isolated.
Instead the electron shell of phosphorus is enlarged by addition of an electrophile yielding a λ^5-phosphorine derivative.
λ^5-phosphorins constitute a novel and very versatile class of heterocyclic compounds.
λ^3- and especially λ^5-phosphorins are electron-rich aromatic compounds, comparable with aniline.
(123 references)

Springer-Verlag
Berlin
Heidelberg
New York

München Johannesburg
London New Delhi Paris
Rio de Janeiro Sydney
Tokyo Wien

Volume 41

New Concepts 1

31 figures. IV, 152 pages
Cloth DM 48,—; US $19.70
ISBN 3-540-06333-1

Contents: B. M. Trost, Sulfuranes in Organic Reactions and Synthesis; W. Kutzelnigg, Electron Correlation and Electron Pair Theories; R. G. Pearson, Orbital Symmetry Rules for Inorganic Reactions from Perturbation Theory; H. Gelernter; N. S. Sridharan; A. J. Hart; S. C. Yen; F. W. Fowler; H. Shue, The Discovery of Organic Synthetic Routes by Computer.

Volume 42

New Concepts 2

Approx. 93 figures. Approx. 160 pages. 1973
Cloth DM 54,—; US $22.20
ISBN 3-540-06399-4

Contents: M. Simonetta, Qualitative and Semiquantitative Evaluation of Reaction Paths; I. Gutman; N. Trinajstić, Graph Theory and Molecular Orbitals; E. Scrocco; J. Tomasi, The Electrostatic Molecular Potential as a Tool for the Interpretation of Molecular Properties.

Volume 43

New Concepts 3

Approx. 19 figures. Approx. 120 pages. 1973
Cloth DM 48,—; US $19.70
ISBN 3-540-06400-1

Contents: P. Čarsky; R. Zahradník, MO Approach to Electronic Spectra of Radicals; H. Hartmann; K.-H. Lebert; K.-P. Wanczek, Ion-Cyclotron Resonance Spectroscopy.

Prices are subject to change without notice